THE FOUNDERS OF EVOLUTIONARY GENETICS

BOSTON STUDIES IN THE PHILOSOPHY OF SCIENCE

Editor

ROBERT S. COHEN, *Boston University*

VOLUME 142

THE FOUNDERS OF EVOLUTIONARY GENETICS

A Centenary Reappraisal

Edited by

SAHOTRA SARKAR

Department of Philosophy and Theoretical Biology Group,
Boston University

KLUWER ACADEMIC PUBLISHERS
DORDRECHT / BOSTON / LONDON

Library of Congress Cataloging-in-Publication Data

```
The Founders of evolutionary genetics : a centenary reappraisal /
edited by Sahotra Sarkar.
       p.   cm.  --  (Boston studies in the philosophy of science ; v.
142)
    Includes bibliographical references and index.
    ISBN 0-7923-1777-7 (HB : acid-free paper)
    1. Evolution (Biology)  2. Genetics.  3. Ecological genetics.
I. Sarkar, Sahotra.  II. Series.
Q174.B67  vol. 142
[QH371]
001'.01 s--dc20
[575.1]                                                    92-12823
```

ISBN 0792317777 (hb)
ISBN 0792333926 (pb)

Published by Kluwer Academic Publishers,
P.O. Box 17, 3300 AA Dordrecht, The Netherlands.

Kluwer Academic Publishers incorporates
the publishing programmes of
D. Reidel, Martinus Nijhoff, Dr W. Junk and MTP Press.

Sold and distributed in the U.S.A. and Canada
by Kluwer Academic Publishers,
101 Philip Drive, Norwell, MA 02061, U.S.A.

In all other countries, sold and distributed
by Kluwer Academic Publishers Group,
P.O. Box 322, 3300 AH Dordrecht, The Netherlands.

Printed on acid-free paper

Printed in The Netherlands

TABLE OF CONTENTS

SAHOTRA SARKAR

THE FOUNDERS OF THEORETICAL EVOLUTIONARY GENETICS: EDITOR'S INTRODUCTION

1

The primary concern of this volume is the role of four individuals, R. A. Fisher (1890—1962), J. B. S. Haldane (1892—1964), H. J. Muller (1890—1967) and S. Wright (1889—1988) in the history of theoretical "evolutionary genetics," though their other contributions also receive significant critical attention. The term, "evolutionary genetics" is construed here more broadly than "population genetics," or the genetic analysis, theoretical and experimental, of the properties of populations subject to various processes such as mutation, selection and drift. At the theoretical level the foundations of population genetics were laid down by Fisher, Haldane and Wright in the 1920's and it is customary to consider their work together. Working independently, during that period, the three of them systematically explored the mathematical consequences of Mendelian inheritance. Even earlier, in 1918, Fisher had managed to derive the regularities reported by the biometricians — Galton, Pearson, Weldon and their collaborators —, including the Law of Ancestral Heredity, from the assumptions of Mendelian inheritance (Fisher 1918). It was a straightforward case of theoretical reduction, of biometry to Mendelism though, somewhat mystifyingly, it has often enough been called a "synthesis" (see Sarkar 1993).

During the 1920's Fisher and Wright developed what became fully general, but oddly univocal, "theories" of evolution based on genetics. Fisher put his faith entirely on natural selection even though, in 1922, in a paper whose inappropriate title ('On the Dominance Ratio') should have established him as a humorist, he had been the first to realize that random processes could be critical to the survival of an allele in a population (Fisher 1922a). Wright, during this period, emerged with the "shifting-balance" theory of evolution, in the first versions of which (Wright 1931, 1932), a myriad of factors had to co-operate almost miraculously for evolution to have taken place.

Haldane, meanwhile, was concerned with more mundane matters.

1

He worked out, in excruciating detail, with great mathematical precision and very little elegance, the exact consequences of the operation of selection in a wide variety of circumstances. Though by the end of the decade Haldane was espousing some views similar to Wright's, he presented no univocal theory of evolution. Much more than Fisher and Wright, he was concerned with connecting experiment with theory. In 1924, for example, he predicted from one of his models, on the basis of very meager field data, that the fitness of the dominant melanic form of the peppered moth was 50% greater than that of the recessives in the Manchester environment (Haldance 1924). At that time this selection intensity was considered far too high to be probable; thirty years later, the field work of Kettlewell (1956) would show it to be quite reasonable.

Haldane's concern with evidence naturally led him to consider the relation of Mendelism, an abstract theory where genes could just as well be stored in bank vaults while determining the properties of future generations (somewhat like inherited wealth), to the physical process of heredity. In the early 1920's this led automatically to Columbia University, to its fly room, and to the Morgan group. In the *Causes of Evolution* (1932), where Haldane presents the first systematic summary of everything that he considered to be relevant to the genetic basis for evolutionary change, an entire chapter is devoted to the model of the chromosome elucidated by the Morgan group.

Arguably, the only genuine theorist in the Morgan group was Muller. Also arguably, without the Morgan group's work, neither Haldane nor anybody else could have begun to connect theoretical population genetics with much potentially quantitative evidence. If one looks at the historical origins of the theoretical understanding of population genetics, then, one must look also at Muller along with Fisher, Haldane and Wright. But Muller was no mathematician and the style and substance of most of his work is far from what conventionally goes as theoretical population genetics. To regard him as a founder of that discipline would, therefore, be irrelevantly controversial. Instead, the term "evolutionary genetics" is being expropriated in this volume to consider all four figures — Fisher, Haldane, Muller and Wright — together. The rationale behind the use of this term is to find a relatively neutral way to consider together all the *theoretical* work that went into the demarcation of the types of theories of evolution that genetics allows.

This is not even remotely the first use of the term "evolutionary genetics." It is simply the appropriation of that term, very likely with insufficient knowledge and respect for its past usage. For that, the Editor alone is responsible and requests tolerance. He has, as far as he can tell, no intention or desire to use it for any historiographical purposes other than that just mentioned. Even more important, the decision to consider Muller together with Fisher, Haldane and Wright is also not original. Crow (1984) has already done so, arguing persuasively that Muller was "keenly interested in evolution and made substantial contributions to the development of the neo-Darwinian view." Crow's reasons for considering these four figures together and the reasons discussed above are complementary. This book continues a historiographical choice he initiated; others will have to judge whether it is appropriate.

The foregoing considerations were intended to show why Fisher, Haldane, Muller and Wright should be considered together in the history of theoretical evolutionary genetics.[1] By a welcome stroke of luck, from the point of view of the Editor, all four of these figures were born almost together, between 1889 and 1892, and almost exactly a century ago. It therefore seemed appropriate to use their birth centenaries to consider their work together. A conference was held at Boston University, on March 6, 1990, under the auspices of the Boston Center for the Philosophy and History of Science, to discuss their work. This book has emerged mainly from that conference. Maynard Smith was invited to speak but, unfortunately, could not participate. Sarkar was present but did not deliver a paper. Provine's contribution to this volume is different from the paper that he delivered at the conference where he spoke of Haldane's role in Kimura's formulation of the neutral theory of evolution. Moreover, John Turner participated in the conference but has not contributed to this volume.

Beside being a report on that conference, this volume also attempts to provide a relatively complete centenary reappraisal of the work of the founders of evolutionary genetics. Hence, some particularly pertinent previously published papers have been included. Moreover, for the sake of completeness, for each of these figures, appraisals of their work in fields other than theoretical evolutionary genetics have been included wherever possible. The organization of this volume reflects the Editor's rigid and unthinking devotion to the Latin alphabet for want of any obvious better ordering principle. Fisher, Haldane, Muller and Wright are considered in that order. When a paper considers more than

one of these figures, it is placed immediately after both figures have
been introduced. The rest of this Introduction is an attempt to situate
the various papers in their intellectual context and to highlight some of
their contents.

2

Even if Fisher's work turns out ultimately to have no biological significance, which is hardly likely, Fisher would still remain one of the
twentieth century's major scientific figures simply because of his work
in statistics. No assessment of Fisher as a scientist is complete without
at least some consideration of his statistics. Moreover, sometimes, it is
impossible to separate Fisher's contributions to biology from his contributions to statistics. Note, for example, that the concept of variance,
and the analysis of variance, were first introduced by Fisher (1918) in
a biological context in his famous "Correlation of Relatives" paper
mentioned above. Fisher was not unique in this respect. In biology, his
most similar conceptual predecessor was Pearson whose statistical
work, along with the earlier largely intuitive statistical methods developed by Galton, was also generally done in an evolutionary context.

Seidenfeld's contribution to this volume is an assessment of a
particularly interesting aspect of Fisher's statistical work: the relation
between two of his most influential contributions in statistics, namely,
his theory of estimation and his theory of experimental design. Seidenfeld argues that the connection between these two theories is achieved
through Fisher's (technical) account of Information: improvements of
experimental design are to be quantified by the increase in the Information provided by the new statistical estimates that can then be derived
from the experimental data.

As in his theory of evolution (see Fisher 1922a), Fisher's ground-
breaking work in the theory of estimation goes back to the early 1920's
(Fisher 1922b, 1925) when he provided formal criteria for the adequacy of a statistical estimator. One of these criteria, more important
than mere consistency, is the amount of Information it provides.
Seidenfeld uses the χ^2-significance test to illustrate that this criterion
can be used to justify maximum likelihood estimation. Fisher's work on
the design of experiments starts a little later (Fisher 1935) but, as
Seidenfeld argues, he is explicit that the improvement of an experimental design results in the data obtained carrying more Information.

Yet, this principle, as Seidenfeld also argues, does not offer any rationale for randomization in experimental design, a strategy that Fisher also strongly advocated. Seidenfeld leaves this puzzle unresolved but its very existence illustrates that it is hardly surprising that randomness in experimental design remains controversial as a methodological principle even today.

3

For almost ten years John Maynard Smith worked with Haldane at University College, London. As an undergraduate, Maynard Smith had studied engineering at Cambridge.[2] In the late 1930's, like Haldane and many others, he had joined the Communist Party primarily because the communists were politically unique in offering principled resistance to the rise of nazism and fascism. He had long known of Haldane's work since his student days at Eton, when he had read *Possible Worlds* (Haldane 1928). While at Cambridge he had invited Haldane to address the university Socialist Club, of which he was then the secretary. Haldane had agreed, but a week before the scheduled event, insisted that he could come only if a train back to London could be guaranteed for the same night. This was during the war; there was bombing going on and no train at night. Faced with an audience and a hall, Maynard Smith nevertheless sent Haldane a telegram assuring him that a train would be available. Haldane found out to the contrary only upon arrival. Irate, he demanded a bicycle; at the end of his lecture he was going to cycle back to London. However, the success of the lecture mollified him and, afterwards, he went quietly to his hotel.

Perhaps luckily for Maynard Smith, Haldane apparently did not remember him from this incident when he began to study at University College, London, in 1948. Maynard Smith had spent the previous seven years at Reading as a technician working for a small aircraft factory. During these years he had decided to turn to biology as a career. On October 1, 1947 he wrote to Haldane. "Dear Comrade," the letter began, "I am writing to ask you if you can give me any advice, as I have decided to take a degree in biology with a view to doing research. I am at present 27 years old, and am employed as an aircraft technician . . . [but] my interests are far more in the direction of science than engineering, and in particular I am interested in biology." The letter concludes: "As an explanation of the 'comrade,' I am at present

Secretary of the Reading Branch of the Party — a fact for which you are partly responsible." Haldane wrote back on October 3rd, mentioning that only if Maynard Smith "did Zoology would [he] come in contact with [Haldane] to any great extent." Not surprisingly, Maynard Smith opted for that subject.[3]

Thus began one of this century's most fascinating and influential biological careers. Maynard Smith and Haldane remained friends until the latter's death. Ironically, perhaps, they never wrote a paper together. Maynard Smith's contribution to this volume is not only a very informed assessment of Haldane's work, it will also undoubtedly rank as an important personal document that future historians can use as a primary source for the history of evolutionary thinking in this century.[4] Two of Maynard Smith's personal appraisals are of special interest. The first is his analysis of the role played by Marxist thinking in the work of two of the most influential of communist biologists: Haldane and himself. From Maynard Smith's point of view the former was less explicitly influenced by Marxist theory, in fact, so little that Marxism seems to have been largely irrelevant. This is an unexpected conclusion given the extent to which Haldane wrote about Marx and Engels. Independent of Maynard Smith, the Editor would like to endorse, strongly, this conclusion.

The second particularly important personal appraisal is that of the dilemma posed by the Lysenko affair to all communist biologists. Maynard Smith notes that if one assumed that the Marxist dialectic is a good guide for science, one could prejudiced towards the inheritance of acquired characteristics (the ability of the soma to influence the germ-line). In fact, he felt such an influence himself. Pointedly, he does not remember Haldane ever making such an argument. That Haldane took Lysenkoism seriously as a scientific possibility was a result of his open-mindedness and, as Maynard Smith notes, his belief that one should take seriously any alternative to one's favorite orthodoxy. In *Evolutionary Genetics* (1989), Maynard Smith also endorses the same methodological principle.

Maynard Smith's account of Haldane's work is primarily about his work in evolutionary biology though the work on enzyme kinetics and the origin of life is mentioned. This analysis reveals a historical fact of some significance: around 1930 Haldane was offering what amounts to a version of Wright's shifting balance theory. In fact, Haldane (1931) anticipates parts of that theory and the *Causes of Evolution* (1932), written before he was aware of Wright (1931), expresses ideas very similar to those soon to be championed for decades by Wright. By 1959, as Crow *et al.* (1990) note, Haldane (1959) had come to have some serious doubts. Moreover, Maynard Smith notes that what he remembers from Haldane's lectures in the early 1950's already appeared to be very different from what Haldane (1931, 1932) had endorsed. The causes of this change in view are yet to be fully understood.

Maynard Smith also points out several of Haldane's ideas that were ahead of their time. Among these are his analysis of kin selection and his anticipation of the notion of complementarity in gene replication. Perhaps even more interesting is the possibility that Haldane might have discovered chaotic dynamics in ecology in 1950 had he had access (and was willing to use) digital computers. Elsewhere Maynard Smith (1985) has also pointed out that Haldane (1949) had been prescient enough to note that resistance to disease might be an important factor in evolution. Indeed, Haldane was the first to suggest that malaria might be a cause of hemoglobin polymorphism.[5] The role of disease in evolution has become a topic of intense study during the last decade (see, e.g., Howard (1991) and Sarkar (1992)). Maynard Smith's paper is ample testimony to Haldane's remarkable prescience.

4

Haldane spent the decade 1923—1924 at Cambridge. During this period his most important work, on the mathematical theory of natural and artificial selection, was published, culminating in the highly influential *Causes of Evolution* (1932). Haldane's professional appointment at Cambridge was as Reader in Biochemistry, second-in-command to F. G. Hopkins, at the newly-founded Biochemical Laboratory. This was surely a curious position for an evolutionary geneticist to fill: what were Haldane's contributions to biochemistry? Very little historical work has been done on this topic though some, such as Olby (1974), have summarily dismissed the importance of Haldane's work.

Sarkar's paper in this volume describes Haldane's work and argues, implicitly, against assessments such as those of Olby. Haldane's biochemical work is divided into two parts: (i) the mathematical treatment of enzyme kinetics; and (ii) work on biochemical genetics that Haldane initiated and was carried out by R. Scott-Moncrieff, W. J. C. Lawrence and, later, J. R. Price at the John Innes Horticultural Institute. The work on enzyme kinetics turns out to be of fundamental importance: Briggs and Haldane (1925) initiated the shift from equilibrium to steady-sate assumptions about the interaction of an enzyme and its substrate. Though the new method is standard now, Briggs and Haldane often do not get mentioned. Michaelis and Menten (1913) are the ones usually credited with creating enzyme kinetics though they had only studied it under equilibrium assumptions and, indeed, had been almost fully anticipated by Henri (1903) and Brown (1902).

Sarkar finds Haldane's methodology in enzyme kinetics to be identical to that in his mathematical treatment of selection. A simple system is analytically described, and then explored in detail. Complicating factors are afterwards systematically introduced as perturbations of the simplified model. Sarkar also considers Haldane's role in chemical genetics. Haldane had, indeed, suggested the "one gene-one enzyme" hypothesis long before it became fashionable. All the same, he did not pursue it as vigorously as he could have and, moreover, he received whatever credit was due to him (see, e.g., Beadle and Tatum (1941) where they refer to Haldane, Troland and Wright as several authors who had suggested the "possibility that genes act through the mediation of enzymes").

Sarkar's paper is lacking at least in one respect. Haldane's work on physiology during this period and his paper on the origin of life (Haldane 1929) are ignored. Without at least some cursory treatment of this work, any account of Haldane's biochemistry is incomplete. However, Sarkar's paper forms part of a projected larger work, an intellectual biography of Haldane (to be published by Oxford University Press). Presumably, that work will include a discussion of Haldane's contributions to these areas.

5

Not all theoretical work is necessarily mathematical. Otherwise, Muller simply could not be considered along with Fisher, Haldane and Wright as one of the founders of theoretical evolutionary genetics. Muller was no mathematician. Yet he contributed, sometimes alone, and sometimes in parallel with the other three, many of the *theoretical* ideas and principles that evolutionary genetics draws upon. Many of these are so commonplace now that it is often not realized how novel, and important, they were when first articulated. Consider a few examples, all drawn from Crow's contribution on Muller in this volume: (i) Muller (and Bridges) suggested that genes do not arise *de novo*. New genes arise out of mutations of old ones; (ii) though preceded by Haldane (1937) by thirteen years, Muller (1950b) independently introduced the concept of the mutation load; (iii) in the 1930's Muller suggested that species differences are chromosomal though here, too, he had been preceded by Haldane (1932); (iv) Muller (1950a) went at length to point out that extreme precision of selection was a hallmark of evolution; and (v) in the 1920's Muller (e.g., Muller 1922) advocated the view that the gene is the basis for life, thereby foreshadowing many of the ideas such as that of the selfish gene that are so fashionable nowadays.

Many of these ideas no doubt require modification in view of developments since, in both biology and philosophy. However, it is hard to argue either against the importance of these ideas of that they are not properly regarded as part of the theoretical or conceptual framework of evolutionary genetics. These and many other of Muller's contributions to evolutionary biology are systematically discussed in the first paper by Crow in this volume. Crow is ideally situated to make a proper assessment of Muller's role in evolutionary biology. Not only is he a leading geneticist, Crow was also one of Muller's collaborators and it is his initiative, as has been noted above, that led to the historiographical choice of including Muller with Fisher, Haldane and Wright among the founders of theoretical evolutionary genetics. Crow's paper can be taken as providing the rationale for that inclusion.

In particular, Crow treats Muller's ideas for the origin and evolution of sex in detail. Muller (1922) and Fisher (1930) are responsible for developing the idea that the advantage of sex lies in the ability of recombination to combine two already favorable mutations that had arisen by chance in different strains.[6] Others, including Haldane (1924) had partly anticipated the insight, but had not developed it. Muller himself made his ideas more quantitative in 1958 (Muller 1958). Muller's 1932 paper made a strong impression on Crow who also spent 1959 as a visiting teacher at Indiana University where Muller was then situated.[7] Muller attended Crow's population genetics lectures and frequent discussion ensued. By 1964 Crow had found an error in one

of Muller's ideas from 1958 the "competition effect," as Muller acknowledged in a postcard (August 22, 1964).[8] Muller's quantitative
treatment was subsequently significantly improved by Crow and Kimura
(1965). Muller's error was not mentioned because he had apparently
intended to write more about the evolution of sex and Crow thought
that Muller would prefer to correct his own error. Unfortunately,
Muller died too soon to do so.

Though no longer regarded as being the most likely mechanism for
the maintenance of sex today, as Crow notes, the Muller—Fisher
principle was highly influential in its time and might have been the main
factor responsible for the survival of sex in an earlier phase of evolution. What this aspect of Muller's work here demonstrates, if more
argument is necessary, is the extent to which Muller's theoretical work
intersected with that of Haldane and Fisher. In the case of Muller's
relation to Haldane, Wimsatt's paper in this volume provides even more
proof of a confluence of interests.

Crow also notes Muller's work in human genetics and his lifelong
commitment to eugenics, not only negative eugenics (the attempt to
eliminate obviously deleterious alleles from the population), but also
positive genetics, that is, the attempt to breed for desirable traits.
Muller's views were controversial even during his life and the possibility
of positive eugenics, as social policy, hardly seems any more likely now.
However, with the identification of many diseases with a clear genetic
basis, negative eugenics, through amniocentesis and selective abortion,
is already being practiced and, except in those societies that are suffering from Christian and similar fundamentalisms, will only be practiced
even more. Moreover, the sequencing of the human genome and further
research into behavioral genetics, whatever its quality, will increase the
political prospects for some positive eugenics. Those who argue against
positive eugenics almost as a matter of principle — as did Haldane who
loved to emphasize the limitations of the present human knowledge of
what is truly desirable — might have to address Muller's arguments.

Crow also provides a candid assessment of the differences between
the four figures who form the subject of this volume with respect to the
generality of their interests, their scientific style, and their mathematical
tastes. Not surprisingly, Fisher emerges as being mathematically the
most sophisticated. The dominance of the shifting balance theory in
Wright's thinking is emphasized. Compared to Fisher, Haldane's and
Wright's mathematical work is found lacking in elegance. Muller was

even less mathematical but none except Haldane could surpass him in the breadth of scientific interest.

6

Linkage mapping, or the establishment of the order of various loci along a chromosome was critical in the establishment of the linear model of the chromosome early in this century. It has become critical again in the current attempts to locate various traits including, very controversially, complicated human behavioral traits to particular sites on chromosomes. The early history of linkage analysis was as fraught with controversy as its use is today. The idea of linkage, or the tendency of certain alleles to be inherited together, in violation of the Mendelian principle of independent assortment, had itself to fend off the claims of a quite recalcitrant pretender: Bateson and Punnet's theory of "reduplication" which attempted to explain such deviations by differential division rates of cells with different combinations of alleles. Later, the linear interpretation of linkage data was challenged for a while by Castle. Both linkage and the linear model of the chromosome emerged triumphant from these tribulations largely through the work of the Morgan school whose most spectacular member was Muller.

The Morgan school systematically developed what, in hindsight, would be called semi-empirical methods for using recombination frequencies to construct linkage maps. Meanwhile, Haldane (1919) developed a set of highly idealized formulae to carry out that construction. Almost mysteriously, the Morgan school ignored this work completely until they criticized it in 1922 as irrelevant (Morgan et al. 1922). Wimsatt's paper in this volume examines this episode in detail and explicates the differences in research strategy followed by Haldane and the Morgan school, especially Muller. Muller emerges as theorizing from the trenches, fully cognizant of the full complexities of the mechanisms of recombination whereas Haldane, by comparison, appears to be searching for global generalization.

This is a somewhat surprising but revealing characterization of Haldane's research strategy. Compared to Wright and Fisher, Haldane almost always appears to shun grand generalizations. He never presented any single hypothesis as *the* mechanism of evolutionary change and his most important early work in evolutionary biology was the detailed calculation of the effects of selection in a wide variety of cases, taking into account as many complicating factors as possible. Yet, when compared with a self-consciously experimental group, Haldane emerges as a theorist after all. Perhaps, though, this should not appear surprising given Haldane's legendary inability to cope with equipment.

Wimsatt's paper also reveals many novel historical points of immediate interest: (i) what is usually referred to as the Haldane mapping function is not actually what Haldane presented as a mapping function in 1919; (ii) in fact, the "Haldane mapping function" was inspired by an earlier paper by Trow (1913); (iii) that Muller (1916) clearly anticipated Fisher (1949) by over thirty years in advocating the use of contrary crosses in genetics to increase the reliability of results; and (iv) Haldane's 1919 paper seems to be written with an even more than characteristic carelessness and a very uncharacteristic unawareness of the current literature.

The last point is susceptible to easy historical explanation. Haldane had been concerned with linkage (or, rather, "reduplication") very early, and in 1912, he had been the first to discover, though not publish, a case of linkage in a vertebrate (Haldane *et al.* 1915). His scientific career, however, was interrupted by World War I. While convalescing in Delhi during the 1917—18 winter, after being wounded in Iraq, he had done, what he later called, "a little rather second-rate theoretical work in genetics, working on results obtained by Morgan and his colleagues in New York (Haldane *ca.* 1942)." The paper on linkage (Haldane 1919) is presumably a report of this work. Little wonder, then, that Haldane's knowledge of the extant literature was scanty.

Wimsatt's paper in this volume is part of a larger project on the history of linkage mapping and a philosophical assessment of model-building in that area. It is exactly the kind of detailed, and conceptually sophisticated, intellectual history that philosophers find valuable when attempting to grope with the nature of scientific change. Though historical work is rampant in evolutionary biology, this kind of detail and sophistication has been sadly lacking and much more work along these lines has long been overdue.

7

Crow's second contribution to this volume is his assessment of Wright's place in the history of twentieth-century biology.[9] Though Crow and Wright were closely associated as colleagues for over three decades, this paper is not overtly autobiographical. However, it is perhaps because of their association that it provides a brilliant and very complete introduction to Wright's work. Crow particularly notes Wright's selfless service to others, as a referee and commentator. Some examples are given here; one other, from a more autobiographical piece (Crow 1982) is perhaps even more illustrative. In the early 1950's, Crow and Newton Morton, both at Madison, were involved in experimental studies of Wright's concept of effective population number. Requiring some mathematical assistance, they visited Wright, at the University of Chicago, with a carefully prepared list of about 12 questions. To each of them, Wright gave the same response: he did not know the answer. The meeting was over in fifteen minutes. On the way back to Madison, Morton noted that if Wright was the greatest population geneticist alive, the subject was in bad shape. A week later, however, Crow and Morton received a 14-page letter from Wright answering each question in full detail. Wright was not prone to express snap judgments on complicated problems but he took whatever time that was necessary to help others with their research.

Crow discusses Wright's work in statistics, animal breeding, physiological genetics, population genetics, the shifting-balance theory of evolution, and philosophy in detail. This paper is easily the best introduction to Wright's work that is available; for those interested in more detail, Provine's biography of Wright (Provine 1986) is strongly recommended. Crow concentrates most on the shifting-balance theory of evolution and, in the process, provides a very clear and non-technical exposition of that theory which goes back to the earliest period of Wright's work, to the 1920's. Wright (1932) argued that the effect of selection on a population could be described by a climbing path on a surface whose height represented fitness and each point on which corresponded to the genotype that was dominant in a given population.[10] If selection is the only mechanism operative, each population would get stuck at the first local peak or maximum of fitness in which it would find itself. Further evolution would be impossible, since to get within the domain of attraction of any other peak would require the population to pass through valleys of lower fitness in the intermediate stages.

Wright attempted to resolve this dilemma by espousing his three-phase "shifting-balance" theory of evolution (perhaps most succinctly described in Wright (1965)). A population is assumed to be composed of many partially isolated local (sub)populations. In the first phase, these cross many of the valleys by random processes, such as genetic drift, while establishing these many local populations. In the second phase, mass selection drives populations to higher peaks. In the third phase, individuals from populations at the highest peaks, that is, individuals with the currently optimal genotypes, migrate to other populations and, through selection, increase the general fitness of these populations. Thus evolution is a result of a balance of all these factors, not just selection.

Wright developed the first part of the theory in great mathematical detail. The second part only requires trivial mathematics. He, however, left his discussion of the third part entirely qualitative and it is this part that many critics, such as Haldane (1959), found problematic mainly because recombination could potentially break up the favorable combination of alleles in the rare migrants to less fit populations. Recently, however, Crow et al. (1990) have begun simulations of the third phase. Results obtained so far, for 2—9 loci in which all the intermediate genotypes between two extreme ones are equally unfit, support Wright's intuition even with low migration rates. It does not even matter whether migration is unidirectional, that is, only from populations with higher fitness to those with lower.

Crow mentions these results and discusses the empirical status of the shifting-balance theory. He argues that the neutral theory (Kimura 1983) can also be interpreted to support Wright. Over all, his assessment of the theory is largely positive. This is important because, ultimately, Wright's place in evolutionary biology is very likely to depend on the outcome of the shifting-balance theory since he has come to be so closely identified to it.

8

The last two papers in this volume, by Provine and Hodge, are both comparative studies of Fisher and Wright. They complement each other almost perfectly. Provine compares and contrasts Fisher's and Wright's evolutionary theories; Hodge deals with that only very briefly but takes up, in detail, their phiosophical and ideological views.[11] Provine's contribution is a detailed comparison of Fisher's and Wright's work on evolution. Fisher and Wright strongly disagreed on many specific evolutionary questions including, thought not limited to, the evolution of dominance and the causes of polymorphism in *Panaxia dominula*, the two cases that Provine analyzes in detail. Provine shows that these disagreements are even more profound than usually thought: they stem from radically incompatible views of the general mechanism of evolution in nature.

The difference, here, has nothing to do with their widely-different mathematical techniques in population genetics. Fisher used all the usual techniques of mathematical physics with which he was fully familiar by training. Wright invented novel, and sometimes almost unbelievably cumbersome, techniques such as path analysis as he went along. Yet, all numerical discrepancies between Fisher's and Wright's work were resolved without difficulty. As Provine points out, where Fisher and Wright disagreed was at the qualitative level: what are the most *important* mechanisms of evolution in nature? Fisher was probably the strongest selectionist that evolutionary theory has ever known: evolution proceeded by constant and continuous mass selection on single genes in large panmictic populations over a long period of time. Wright preferred his shifting-balance theory. Though he did not deny the importance of selection, at least not always, population structure, drift and isolation were equally necessary for evolution.

It would appear that this difference could be settled by data and Provine argues that this is precisely what the experimentalists, that is, the naturalists such as Dobzhansky and Ford set out to do. As Crow's paper on Wright points out, the verdict is still not in. The future might decide in favor of Fisher or Wright or perhaps neither, if there is no single dominant mechanism of evolutionary change or if such a mechanism, when established, turns out to be radically different from what Fisher or Wright had championed. It is instructive to note, in this context, that Haldane was never convinced by Fisher or Wright. Though obviously more impressed by the power of mass selection than Wright, he saw no reason to postulate any single mechanism as the cause of evolution. The historical process of evolution, Haldane seems to suggest, is far too complex. Each case of evolution can be analyzed mathematically. But evolution and, especially, natural selection has to be modeled case by case. It might turn out that evolution is as complex as Haldane thought, driven by different mechanisms in different contexts. If this turns out to be the eventual outcome of future research, it would be a curious end to the dispute between Fisher and Wright, a dispute which, as Provine argues, has so strongly influenced the nature of evolutionary theorizing in the last few decades.

9

Hodge's paper offers another comparison of Fisher and Wright, not only about the substance of their evolutionary theories, but more generally, about their respective philosophical views and ideological preferences. Such comparisons are always interesting to all historians and philosophers for science except perhaps those who are interested only in "conceptual history" in the narrowest possible construal of that term. However, in the cases of Fisher and Wright, two additional factors make this comparison particularly interesting. *First*, though amateurs, both Fisher and Wright wrote quite extensively on philosophical themes (as, indeed, did Haldane). *Second*, in Fisher's case it has long been recognized that a deep commitment to extreme conservatism, including radical eugenics, played an important role in his evolutionary theorizing. Indeed, the last five chapters of the *Genetical Theory* have little value outside the range of the standard eugenic concerns of that time. In contrast, Wright's consistent failure to take any significant stands in favor of eugenics is suggestive exactly in that context.

Hodge's paper is revealing both with respect to both Fisher and Wright, and with respect to some general historiographical issues about twentieth-century evolutionary theory, in particular, the question whether this history is to be interpreted as a triumph of dialectical or mechanistic materialism. Hodge's treats all these matters in requisite detail. His treatment defies summarizing: here, only two particularly intriguing points will be mentioned.

Though Fisher's debt to physics is well-known, Hodge attempts to give a more systematic account of the influence of statistical mechanics and the kinetic theory of gases on Fisher than has previously been attempted. In this context, Hodge argues for a very interesting position, namely, that Fisher was prone to giving an *ontic* interpretation of indeterminism in statistical reasoning in classical physics, that is, in the context of the work of Boltzmann, Gibbs and Jeans. This indeterminism, according to Hodge's interpretation of Fisher, was in the world itself and provided a basis for further indeterminism. All quantum

mechanics does is simply reinforce this conclusion rather than introduce a new sort of indeterminism in physics. *If* Hodge is correct, and
the evidence he presents seems quite compelling, then *either* Fisher was
guilty of a remarkable conflation of ontological and epistemological
issues *or* the standard philosophical interpretation of classical statistical
mechanics, which invokes only an epistemic interpretation of indeterminism is incorrect. The Editor strongly suspects that the former of
these possibilities is more likely and Hodge deserves commendation for
bringing this puzzling aspect of Fisher's views out in the open.

Wright self-consciously kept his philosophical views, especially his
panpsychism, separate from his science, arguing basically that science
concerned itself only with material aspects of phenomena and could
not, therefore, cast light on other aspects. It has, therefore, been
customary to regard Wright's philosophical views as being irrelevant to
his science. Indeed, this is the strategy followed by his biographer,
Provine (1986). Hodge, however, argues for a deeper connection than
is usually suspected. For example, though Wright was sympathetic to
different units of selection up to the genotypic or individual levels,
according to Hodge, he very rarely had much use for higher levels of
selection. Hodge argues that this could well be because Wright regarded, on the basis of his philosophical views, that up to the level of
the individual organism, all entities have an integrated organization.
Beyond that level, this organization becomes much looser — hence, its
failure to behave as a unit. Hodge uses this aspect of Wright's philosophical views to illustrate some of his other scientific views, including
his emphasis on the possibility that individual adaptability counters the
action of selection. Here, too, Hodge has opened up new problems for
philosophical and historical reassessment.

10

What has been said so far should suggest that, in several ways, the contributions in this volume break new ground in the history of twentieth-
century evolutionary biology. However, this book is hardly the final
word on these topics. There are still many facets of the individual
contributions of Fisher, Haldane, Muller and Wright and their relation
to twentieth-century biology that have yet to be explored:

(i) Of the four, only Wright has yet been the subject of a biography
that considers his work in its full scientific context (Provine 1986). The

existing biographies of Fisher (Box 1978), Haldane (Clark 1969), and Muller (Carlson 1981) are largely personal biographies. In particular, Clark (1969) largely ignores Haldane's scientific achievements and Box (1978) concentrates heavily on Fisher's work in statistics at the expense of his work in evolutionary theory. All three of these biographies need to be supplemented.

(ii) While Wimsatt has begun a philosophical reexamination of the scientific strategies of Haldane and Muller in a limited context, many other questions remain. For instance, what role did their respective ideologies play in their scientific work? In particular, both Muller and Haldane were convinced dialectical materialists for a period. Is there anything in their scientific work that led to his metaphysics? Conversely, how, if at all, did this metaphysics influence their science? Note that Hodge attempts to answer these questions for Fisher and Wright and Maynard Smith provides partial answers in the case of Haldane.

(iii) From the point of view of philosophers of science, what is the methodology followed by Fisher, Wright and Haldane? It has already been suggested (Sarkar 1993), for example, that the relation of biometry to Mendelism, as exemplified in the seminal work of Fisher (1918), was a case of theory reduction. Even if this is correct, does this make the work of Fisher, Haldane and Wright entirely a reductionist enterprise? Note that, in the context of philosophical models of scientific change, very little careful analysis of the emergence of population genetics has yet been presented.

(iv) Fisher, Haldane and Wright worked on the basis of even earlier attempts to explore the mathematical consequences of Mendelism including the work of Detlefsen, Fish, Harris, Jennings, Norton, Pearl, Remick, Robbins, Weinberg, Wentworth and Yule. What, exactly, were these contributions? What was truly original with Fisher, Haldane and Wright? Did they make any conceptual innovations or simply bring more mathematical sophistication to the field? Why should only they, rather than any of the others, be regarded as the founders of theoretical population genetics? Note that Provine (1971) provides partial answers to some of these questions and Sarkar's contribution to this volume attempts to answer similar questions for Haldane's work in biochemistry.

(v) The period during which the foundations of theoretical population genetics were laid down also saw an independent emergence of mathematical ecology. Fisher (1930) and Haldane (1927), at least, were certainly concerned with ecological problems. (The latter even provides a new proof of Lotka's theorem of the stability of the normal age distribution.) It appears, then, that an attempt was being made to develop ecological and evolutionary theory in an integrated fashion. What, precisely, were the conceptual relationships between the new ecology and population genetics? Why, and how, did this integrative attitude subsequently disappear from evolutionary theory?

(vi) It is apparent that more analysis is necessary to provide a full account of the differences between the scientific, ideological and philosophical views of Fisher, Haldane and Wright. Provine's and Hodge's contributions to this volume note some very important differences between Fisher and Wright and Provine (1971) gives some suggestive general answers with respect to all three up to about 1930. However, Maynard Smith notes here how Haldane (1932) seemed similar to Wright; only later does he appear to be different. What caused this change? Where do Haldane and Fisher differ most? To what extent did the differing attitudes of the three to the experimental elucidation of genetic mechanisms by Muller and the Morgan school influence their general view of evolution?

It is not, of course, being suggested that this is even remotely close to a complete list of historical and philosophical questions about either these figures, or about their role in twentieth-century evolutionary biology, that remain to be addressed. In fact, the contributions to this volume point to many other directions for further research. If, even indirectly, this volume leads to the pursuit of any of these topics, it will have served a useful purpose.

11

As the Editor of this volume, it is a pleasure to thank many, besides the contributors, who made the conference and this volume possible. R. Tamarin and R. Hausman had the thankless job of chairing the sessions. T. Lubas and D. Wilkes helped with organization and publicity beyond the call of duty. Oxford University Press generously permitted the reprinting, here, of significant amounts of copyrighted material. And, finally, the Dibner Institute for the History of Science and Technology provided the necessary financial support for the conference thereby relieving what, under the current economic organization of society, might well be the most significant constraint on its intellectual evolution, if not everything else.

Boston University

NOTES

[1] Obviously, Muller has also to be considered in *any* history of biology because of the importance and brilliance of his experimental work in genetics.
[2] Biographical material about Maynard Smith is based on an interview with him, June 12, 1990.
[3] This early correspondence between Haldane and Maynard Smith is preserved in the Haldane Papers (Box 19), University College London Library.
[4] A shorter, and less autobiographical, version of this paper appeared previously in the *Oxford Surveys in Evolutionary Biology* (1986).
[5] See the remarks of Montalenti (in Italian) at the end of Haldane (1949).
[6] Fisher's role in formulating this hypothesis, however, is not very significant as Mooney (1992) has pointed out.
[7] Biographical material about Muller and Crow is based on several conversations with Crow and a recorded interview (March 6, 1990) and a letter from Crow to Susan Mooney (February 1, 1990).
[8] The postcard is to be found in the Muller Mss., Manuscripts Department, Lilly Library, Indiana University.
[9] It is reprinted from the *Journal of the History of Biology* (1990).
[10] Note that Provine (1986) has correctly pointed out that Wright's description is inconsistent because the set of all possible genotypes, over which the fitness surface is to be defined, is a discrete set. Wright, however, only intended fitness surfaces to be used as a metaphor. Crow avoids the problem by constructing the surface over genotypic frequencies which constitute a continuous set. Sarkar (1990) and others have simply modeled Wright's surfaces using only discrete mathematical structures.
[11] Provine's paper is reprinted from the *Oxford Surveys in Evolutionary Biology* (1985).

REFERENCES

Beadle, G. W. and Tatum, E. L. (1941), 'Genetic Control of Biochemical Reactions in Neurospora', *Proceedings of the National Academy of Sciences (USA)* **27**: 499–506.
Box, J. (1978), *R. A. Fisher: the Life of a Scientist*, New York: Wiley.
Briggs, G. E. and Haldane, J. B. S. (1925), 'A Note on the Kinetics of Enzyme Action', *Biochemical Journal* **19**: 338–339.
Brown, A. J. (1902), 'Enzyme Action', *Transactions of the Chemical Society* **81**: 373–388.
Carlson, E. A. (1981), *Genes, Radiation, and Society: the Life and Work of H. J. Muller*, Ithaca: Cornell University Press.

Clark, R. (1969), *J B S — the Life and Work of J. B. S. Haldane*, New York: Coward McCann.

Crow, J. F. (1982), 'Sewall Wright, the Scientist and the Man', *Perspectives in Biology and Medicine* 25: 279—294.

Crow, J. F. (1984), 'The Founders of Population Genetics', in Chakravarti, A., ed., *Human Population Genetics: The Pittsburgh Symposium*, New York: Van Nostrand Reinhold, pp. 177—194.

Crow, J. F. and Kimura, M. (1965), 'Evolution in Sexual and Asexual Populations', *American Naturalist* 99: 439—450.

Crow, J. F., Engels, W. R., and Denniston, C. (1990), 'Phase Three of Wright's Shifting-Balance Theory', *Evolution* 44(2): 223—247.

Fisher, R. A. (1918), 'The Correlation between Relatives on the Supposition of Mendelian Inheritance', *Transactions of the Royal Society of Edinburgh* 52: 399—433.

Fisher, R. A. (1922a), 'On the Dominance Ratio', *Proceedings of the Royal Society of Edinburgh* 42: 321—341.

Fisher, R. A. (1922b), 'On the Mathematical Foundations of Theoretical Statistics', *Philosophical Transactions of the Royal Society (London)* A22: 309—368.

Fisher, R. A. (1925), 'Theory of Statistical Estimations', *Proceedings of the Cambridge Philosophical Society* 22: 700—725.

Fisher, R. A. (1930), *The Genetical Theory of Natural Selection*, Oxford: Clarendon Press.

Fisher, R. A. (1935), *The Design of Experiments*, Edinburgh: Oliver and Boyd.

Fisher, R. A. (1949), *The Design of Experiments*, 3rd ed., Edinburgh: Oliver and Boyd.

Haldane, J. B. S. (1919), 'The Combination of Linkage Values, and the Calculation of Distance between the Loci of Linked Factors', *Journal of Genetics* 8: 299—309.

Haldane, J. B. S. (1924), 'A Mathematical Theory of Natural and Artificial Selection. Part I', *Transactions of the Cambridge Philosophical Society* 23: 19—41.

Haldane, J. B. S. (1927), 'A Mathematical Theory of Natural and Artificial Selection. Part IV', *Proceedings of the Cambridge Philosophical Society* 23: 607—615.

Haldane, J. B. S. (1928), *Possible Worlds* London: Harper and Brothers.

Haldane, J. B. S. (1929), 'The Origin of Life', *Rationalist Annual* 1929: 1—11.

Haldane, J. B. S. (1931), 'A Mathematical Theory of Natural Selection. Part VIII. Metastable Populations', *Proceedings of the Cambridge Philosophical Society* 27: 137—142.

Haldane, J. B. S. (1932), *The Causes of Evolution* London: Longmans, Green and Co.

Haldane, J. B. S. (1937), 'The Effect of Variation on Fitness', *American Naturalist* 71: 337—349.

Haldane, J. B. S. (*ca.* 1942), 'Why I am Cooperator [*sic*]', Unpublished Manuscript, Haldane Collection, University College, London.

Haldane, J. B. S. (1949), 'Disease and Evolution', *La ricerca scientifica, supplemento* 19: 2—11.

Haldane, J. B. S. (1959), 'Natural Selection', in Bell, P. R., ed., *Darwin's Biological Work: Some Aspects Reconsidered*, New York: Wiley, pp. 101—149.

Haldane, J. B. S., Sprunt, A. D., and Haldane, N. M. (1915), 'Reduplication in Mice', *Journal of Genetics* 5: 133—135.

Henri, V. (1903), *Lois Générales de l'Action des Diastases*, Paris: Hermann.

Howard, J. C. (1991), 'Disease and Evolution', *Nature* 352: 565—567.
Kettlewell, H. B. D. (1956), 'A Résumé of Investigations on the Evolution of Melanism in the Lepidoptera', *Proceedings of the Royal Society (London)* B145: 297—303.
Kimura, M. (1983), *The Neutral Theory of Molecular Evolution*, Cambridge: Cambridge University Press.
Maynard Smith, J. (1989), *Evolutionary Genetics*, Oxford: Oxford University Press.
Michaelis, M. and Menten, M. L. (1913), 'Zur Kinetik der Invertinwirkung', *Biochemische Zeitschrift* 49: 333—369.
Mooney, S. (1992), 'The Evolution of Sex: A Historical and Philosophical Analysis', Ph.D. Dissertation, Department of Philosophy, Boston University.
Morgan, T. H., Sturtevant, A. H., Muller, H. J., and Bridges, C. B. (1992), *The Mechanism of Mendelian Inheritance*, 2nd ed., New York: Henry Holt.
Muller, H. J. (1916), 'The Mechanism of Crossing Over. III', *American Naturalist* 50: 350—366.
Muller, H. J. (1922), 'Variation Due to Change in the Individual Gene', *American Naturalist* 56: 32—50.
Muller, H. J. (1950a), 'Evidence of the Precision of Genetic Adaptation', *Harvey Lectures* 18: 165—229.
Muller, H. J. (1950b), 'Our Load of Mutations', *American Journal of Human Genetics* 2: 111—176.
Muller, H. J. (1958), 'Evolution by Mutation', *Bulletin of the American Mathematical Society* 64: 137—160.
Olby, R. C. (1974), *The Path to the Double Helix*, Seattle: University of Washington Press.
Provine, W. (1971), *The Origins of Theoretical Population Genetics*, Chicago: University of Chicago Press.
Provine, W. (1986), *Sewall Wright and Evolutionary Biology*, Chicago: University of Chicago Press, pp. 32—50.
Sarkar, S. (1990), 'On Adaptation: A Reduction of the Kauffman—Levin Model to a Problem in Graph Theory and Its Consequences', *Biology and Philosophy* 5: 127—148.
Sarkar, S. (1992), 'Sex, Disease, and Evolution: Variations on a Theme from J. B. S. Haldane', *BioScience* 42: 448—454.
Sarkar, S. (1993), 'What Fisher Achieved in 1918: the Reduction of Biometry to Mendelism', Forthcoming.
Trow, A. H. (1913), 'Forms of Reduplication — Primary and Secondary', *Journal of Genetics* 2: 313—324.
Wright, S. (1931), 'Evolution in Mendelian Populations', *Genetics* 16: 97—159.
Wright, S. (1932), 'The Roles of Mutation, Inbreeding, Crossbreeding and Selection in Evolution', *Proceedings of the Sixth International Congress of Genetics* 1: 356—366.
Wright, S. (1965), 'Factor Interaction and Linkage in Evolution', *Proceedings of the Royal Society (London)* B162: 80—104.

T. SEIDENFELD

R. A. FISHER ON THE DESIGN OF EXPERIMENTS AND STATISTICAL ESTIMATION

0. INTRODUCTION AND OUTLINE

On this occasion of the centenary year of Fisher's birth, my purpose in this talk is to consider the relation between two of Fisher's major contributions: his theory of experimental design and his theory of statistical estimation. It is no coincidence that each of Fisher's three principal books on statistics ends with a substantial chapter on statistical estimation:

Statistical Methods for Research Workers (SMfRW) 1925 — 14th ed. 1970;

The Design of Experiments (DoE) 1935 — 8th ed. 1966;

Statistical Methods and Scientific Inference (SM&MI) 1956 — 3rd ed. 1973.[1]

The thesis of my presentation is that Fisher linked experimental design and estimation through his technical account of (Fisher-) Information. In particular, improvements in an experimental design, e.g., better controls, blocking, or other factorial restrictions, may be quantified by an increase in the Information provided by estimates derived from the experimental data.

In Section 1, I sketch Fisher's theory of estimation with an eye on explicating the role Information plays in resolving choices of rival estimates derived from a sample. As an illustration of this approach, I show how Fisher's χ^2-significance test for ordinary contingency tables may be decomposed to reveal that Information justifies maximum likelihood estimation. The same example indicates the importance Fisher placed on the *principle of ancillarity*, to wit: conditioning on ancillary data. In Section 2, I discuss the application of Fisher Information to questions of experimental design, and illustrate how added controls (in matched pairs) lead to data with greater Information. The presentation concludes with a brief discussion of one difficulty in this reconstruction of Fisher's use of Information: the problem of randomization. In estimation it is to be avoided (as a result of the ancillarity principle), yet more

23

Sahotra Sarkar (ed.), The Founders of Evolutionary Genetics, 23–36.
© 1992 Kluwer Academic Publishers. Printed in the Netherlands.

than anyone else Fisher is responsible for the theory of randomized experimental design. How is this conflict to be resolved?

1. INFORMATION AND FISHER'S THEORY OF ESTIMATION — WITH AN APPLICATION TO 2 × 2 CONTINGENCY TABLES

In his ground-breaking 1922 and 1925 papers on estimation, Fisher offers formal criteria for determining the adequacy of a statistical estimator. In particular, the arguments he gives press for supremacy of maximum likelihood estimation. In rough form, these and later versions of his theory are organized into a hierarchy of criteria, a sequence of ever finer sieves to distinguish among rival estimates. Those which pass the simpler tests (e.g., for "consistency") are subjected to the heightened scrutiny of more refined tests (e.g., for "efficiency"). But what is the goal of estimation? How are the test criteria justified? The key to answering these questions is that for Fisher an estimate is, first of all, a statistic — a reduction of data. That is, an estimate is appraised as a summary of evidence, not as a "guesstimate" of some unobserved quantity.

The first requirement in estimation is Fisher's criterion of consistency. (Fisher-) Consistency of an estimate identifies the quantity (the parameter) about which the summary is directed. It finds its mature formulation (Fisher, 1956, §6.2) as follows:

DEFINITION. A (*Fisher-*) *Consistent Statistic* is a function of the observed frequencies which takes the exact parametric value when for these frequencies their expectations are substituted.

Suppose, for example, the (i.i.d.) N data are categorial, each occupying one of m cells, with probability p_j for the jth cell ($j = 1, \ldots, m$). Denote the observed cell frequencies by a_j/N, where a_j is the jth cell count. Consider a (linear) statistic A of the form $A = \Sigma_j c_j a_j$, for known constants c_j ($j = 1, \ldots, m$). If we substitute for the cell counts their expected values (Np_j), we obtain an estimate $A/N = \Sigma_j c_j p_j$, which (by definition) is a consistent estimate of this (linear) function of the cell probabilities. For instance, it might be that, as with data relating to the genetic linkage between two characteristics, the parameter of interest, θ, satisfies $\theta^2 = \Sigma_j c_j p_j$. Then $\sqrt{(A/N)}$ is a Fisher-consistent estimate of θ. However, the same data may suggest numerous (Fisher-) consistent estimators for the same parameter of interest.

ILLUSTRATION (SMfRW, §53). Corresponding to estimation of a linkage parameter, $0 < \theta < 1$ ($\theta = 0.5$ for independence between the genes), with four cells having respective probabilities of occurrence on each trial: $\{(2 + \theta^2)/4, (1 - \theta^2)/4, (1 - \theta^2)/4, \theta^2/4\}$, then the following three all are Fisher-consistent estimates of θ.

Estimator$_1$: $\sqrt{[(a_1 + a_4 - a_2 - a_3)/N]}$
Estimator$_2$ (the maximum likelihood estimate): the positive solution to

$$a_1/N(2 + \theta^2) + a_4/N\theta^2 = (a_3 + a_4)/N(1 - \theta^2)$$

Estimator$_3$ (the minimum χ^2 estimate): the positive solution to

$$a_1^2/N^2(2 + \theta^2)^2 + a_4^2/N^2\theta^4 = (a_3^2 + a_4^2)/N^2(1 - \theta^2)^2.$$

What is the reason for insisting on Fisher-consistency of an estimate? The answer lies in Fisher's semantics (that is his "theory") of probability. To assert that the data are an (i.i.d.) sample of N according to the parametrized distribution $p_j(\theta)$ is to require (among other conditions) that the "hypothetical" population from which the sample is taken has cell frequencies given by distribution $p_j(\theta)$ ($j = 1, \ldots, m$). Moreover, the parameter is identified by these population quantities. Thus, a Fisher-consistent estimator for a parameter θ meets the quite minimal condition that, when applied to the population itself, the estimator recovers the quantity θ from such an idealized sample. In that sense, the estimator summarizes the (idealized) evidence of the whole population by the parametric quantity of interest.[2]

The next two, closely related criteria by which estimators are assessed involve Fisher's concept of statistical Information. Formally,

DEFINITION. The (expected) *amount of Information* about θ, $I_\theta[x_1, \ldots, x_N]$, in an i.i.d. sample of N from the distribution $p_j(\theta)$ is

$$-E\left\{\frac{\partial^2}{\partial\theta^2} (\log p_j)\right\}$$

where the expectation is over all possible samples, taken with respect to the distribution $p_j(\theta)$. Information is additive for independent samples, and the Information in a statistic derived from a sample is always bounded above by the Information in the sample as a whole (Fisher, 1925a).

Not surprising, Information serves as a basis in Fisher's theory for discriminating among consistent estimators. Besides consistency, a good estimator T is to be an *efficient* summary of the data from which it is derived. More precisely, with increasing sample size, the ratio of the amount of Information in T_N to the amount of Information in the sample of N from which it is calculated should approach unity:

DEFINITION. Estimator T is (1st order) *efficient* if $\lim_{N \to \infty} I_\theta[T_N] \div I_\theta[x_1, \ldots, x_N] = 1$.

Among the three consistent estimators presented in the gene-linkage illustration (above), the maximum likelihood and the minimum chi-square estimation are efficient. However, the first estimator has an efficiency increasing with the value of the parameter of interest. For example, its efficiency is only about 60% when the two genes are independent ($\theta = 0.5$).

    In order to motivate the final criterion for estimates that I will discuss, consider two statistical extremes with respect to the adequacy of a statistic in summarizing a data set.

DEFINITION. A statistic $T(X)$ calculated from a sample x is *sufficient* for the parameter θ provided that $p(X | T, \theta) = p(X | T)$, independent of θ.

DEFINITION. A statistic $T(X)$ calculated from a sample X is *ancillary* for the parameter θ provided that $p(T | \theta) = p(T)$, independent of θ.

Either by Bayesian or likelihood principles, a sufficient statistic conserves all the relevant evidence (about θ) in the sample from which it is derived, whereas an ancillary statistic is irrelevant to θ. From the standpoint of (Fisher) Information, when $T(X)$ is sufficient for θ, $I_\theta[X] = I_\theta[T]$; T preserves all the relevant evidence in the sample. Likewise, when T is ancillary for θ, $I_\theta[T] = 0$; there is no relevant information contained in an ancillary quantity. Thus, in the case of ancillary quantity T, the statistical analysis may be carried out given T. For example, often it is assumed that sample size, N, is chosen independent of the unknown quantity of interest. Then, N is ancillary and the analysis proceeds with N a known constant.[3]

    Much of Fisher's attention (particularly in his 1934 paper, 'Two New Properties of Mathematical Likelihood') is devoted to establishing the supremacy of maximum likelihood estimation [m.l.e.]. At first (1922) he thought the m.l.e. is sufficient. He weakened his claim (1935, §3) to say that the m.l.e. is sufficient whenever a sufficient statistic exists, and that by conditioning on an ancillary statistic the m.l.e. preserves the greatest quantity of Information that can be summarized in a statistic.[4] In order to see what Fisher had in mind, recall his somewhat controversial treatment of contingency tables.

ILLUSTRATION. For simplicity, let us attend to the elementary case of a 2 × 2 table. Consider data from flips of two coins, summarized in the table below, where we are concerned about the hypothesis that the coins have a common bias.

2 × 2 TABLE

	heads	tails	
coin-1	a	b	$a + b = n_1$ flips
coin-2	c	d	$c + d = n_2$ flips
	$a + c$ heads	$b + d$ tails	$n_1 + n_2 = N$ flips total.

Let θ_0 be an hypothesized value for the common bias of the two coins. Then the χ^2-test for independence (with 2-degrees of freedom) is just the sum of the two, separate 1-degree of freedom χ^2-tests for the two samples (of sizes n_1 and n_2, respectively) about the coins:

$$\chi_2^2 = (a - \theta_0 n_1)^2/\theta_0 n_1 + (b - [1 - \theta_0]n_1)^2/[1 - \theta_0]n_1$$
$$+ (c - \theta_0 n_2)^2/\theta_0 n_2 + (d - [1 - \theta_0]n_2)^2/[1 - \theta_0]n_2.$$

Fisher's controversial proposal for the χ^2-independence test when θ_0 is unknown is to substitute the m.l.e. under the "null" hypothesis, to replace θ_0 with the quantity $(a + c)/N$, resulting in a χ^2 test with 1 degree of freedom. What connection is there between this use of the maximum likelihood estimate in the χ^2 test and the importance that estimates conserve Information? The answer is both subtle and rather surprising.

Significance tests offer a rudimentary form of (Fisherian) statistical analysis against more sophisticated Fisherian tools, e.g., using the likelihood function or, in the special circumstances where it applies, using fiducial inference to relate data to hypotheses. Significance tests are rudimentary in their conclusion — "Either a rare event has occurred or the 'null' hypothesis is false." However, to offset this weakness, Fisher argues that significance tests may be performed even when the space of alternative hypotheses is vaguely specified — unlike the conditions for likelihood or fiducial reasoning.

What makes an outcome "rare"? One scheme for making sense of Fisher's idea (well stated by Cramér, 1946) is to introduce a discrepancy ranking $\mathbf{D}$ on the sample space of possible outcomes Ω, $\mathbf{D}: \Omega \rightarrow \mathfrak{R}$. The intended meaning is that outcomes with higher discrepancy are "rarer" under the null hypothesis. Then, the significance level attained on a trial is the probability (given h_0) of obtaining a discrepancy at least as great as that observed.

With multinomial data, Fisher (SMfRW, §21.02) favored the "exact test," where discrepancy is inversely related to the probability of cell counts. Call this the probabilistic discrepancy ranking, D_p. That is, with m cell counts, $a_j (j = 1, \ldots, m)$, $\Sigma_j a_j = N$, then

$$D_p\{a_j\} = [(N! \div \Pi_j a_j!) \, \Pi_j p_j^{a_j}]^{-1}.$$

Asymptotically, for increasing sample size, the probabilistic discrepancy ranking agrees with the χ^2-discrepancy ranking (on m-1 degrees of freedom), D_{χ^2}; where outcomes are given a D_{χ^2} discrepancy according to their χ^2 values. Hence, with categorical data, χ^2 gives a convenient approximation to Fisher's "exact" significant test.

The controversial aspects of Fisher's treatment of independence tests in contingency tables stems from his claim that the analysis should take the marginal totals as given. That is, Fisher treats the 2 × 2 tables as though the three quantities, N, n_1, and $a + c$ are ancillary.[5] In the illustration with the two biased coins, it is commonplace to assume that the sample sizes (n_1 and n_2, hence also N) are irrelevant to inference from the data. The assertion that $a + c$, too, is uninformative about the null hypothesis (of independence) is without foundation. Clearly, it is false for extreme cases, e.g., when $a + c = 0$.

However, if we grant Fisher's assumption, that the marginal totals give no relevant information in the test of the null hypothesis (that the coins are equally biased), then the "exact" test assumes a convenient form (independent of the value θ_0 of the common bias). Then, given the four marginal totals, $D_p(\{a, b, c, d\}) = [a!b!c!d!]$. How does this compare to the D_{χ^2} discrepancy ranking for the same null hypothesis?

In particular, what of Fisher's use of the m.l.e. in computing the 1-degree of freedom χ^2?

The answer is contained in a decomposition of χ^2. Recall,

$$\chi_2^2 = (a - \theta_0 n_1)^2/\theta_0 n_1 \;\; + \;\; (b - [1 - \theta_0]n_1)^2/[1 - \theta_0]n_1$$
$$+ (c - \theta_0 n_2)^2/\theta_0 n_2 \;\; + \;\; (d - [1 - \theta_0]n_2)^2/[1 - \theta_0]n_2.$$

By some simple algebra,

$$= [\theta_0 - (a + c)/N]^2 \, N \div \theta_0(1 - \theta_0)$$
$$+ (n_1 n_2 \div N)\{a/n_1 - c/n_2\}^2 \div [\theta_0(1 - \theta_0)]. \qquad [*]$$

Write the second summand as $Q^2 = (n_1 n_2)\{a/n_1 - c/n_2\}^2 \div [N\theta_0(1 - \theta_0)]$. It is the negative exponent in the asymptotic (normal) density for the difference in sample "means." That is, asymptotically, a/n_1 and c/n_2 are independently a bivariate normal pair with a/n_1 normal $N(\theta_0, \theta_0(1 - \theta_0) \div n_1)$, and c/n_2 normal $N(\theta_0, \theta_0(1 - \theta_0) \div n_2)$. Hence, the quantity $(a/n_1 - c/n_2)$ is normally distributed $N(0, \theta_0(1 - \theta_0)N \div n_1 n_2)$. Thus, Q^2 provides the "exact," probabilistic discrepancy for a significance test of the hypothesis that the coins are equally biased, using the sample difference in means as the test statistic. By Fisher's assumption that the marginal totals are irrelevant, this statistic exhausts the data.[6]

The "nuisance" parameter (θ_0) appears in Q^2 as part of the variance term; however, it may be removed by "Studentizing" the unknown Normal variance. Here is where the choice of estimate for θ_0 plays a role in the decomposition of χ^2. The left hand summand in [*] is easily recognized as $[\theta_0 - (a + c)/N]^2 \, I_{\theta_0}[N]$, where $I_{\theta_0}[N]$ denotes the Fisher Information about a binomial parameter contained in a sample of size N. Upon adopting an estimate T for θ_0, we see that the left hand summand in [*] becomes: $[T - (a + c)/N]^2 \, N \div T(1 - T)$. This term is positive for all estimates, except when T is the m.l.e for θ_0, that is, except when $T = (a + c) \div N$. Thus, the decomposition [*] of χ^2 provides a way of distinguishing among (1st order) efficient estimators. In short, the left hand term, $[T - (a + c)/N]^2 \, N \div T(1 - T)$, indicates the excess χ^2 discrepancy that results from using an estimator other than the m.l.e. For example, though total χ^2 is reduced by using the minimum χ^2 estimate compared with the m.l.e for θ_0, the minimum χ^2

estimate yields a test of significance with a distorted discrepancy in comparison with the "exact" (asymptotic) test. In fact, Fisher uses a similar decomposition of χ^2 in his discussion of rival (1st order) efficient estimates (SMfRW, §57, especially Figure 12).[7] Thus, we have arrived at one of Fisher's arguments for the supremacy of maximum likelihood estimation, based on Information, understood as a response to the challenge of data reduction without loss of relevant information.

2. INFORMATION AND THE DESIGN OF EXPERIMENTS

2.1. *Control and Precision*

Allow me to begin the second part of my talk with a typical "horror" story about the frustration that statisticians experience when they are called on to consult in "data analysis." A Professor and Graduate student from the Agriculture Division of a noted institution appear at our statistician's door with reams of data from an experiment concerning the (multi-attribute) yield of two varieties of corn. Their data have been carefully collected and recorded. But our satistician is at his wit's end. Alas, test variety 1 of corn was planted on field *A*, test variety 2 of corn was planted on field *B*, and the two fields are not alike! If only the researchers had consulted on the design first. With a few experimental controls their results could have been more revealing even at half the sample sizes!

Fisher [1962] expresses the same problem this way.

When, a little more than 25 years ago, I first attempted a systematic exposition of the subject, known as the Design of Experiments, it is no very grave confession to avow that I did not fully understand the position among the statistical sciences of this new discipline. My approach at that time was frankly a technological one. As a statistician I had often set myself the task of analyzing experimental data, and was much concerned with those improvements in statistical methods which promised to make such analysis, more thorough and more comprehensive. Technically, I could see that some methods were superior to others in the concrete sense of extracting from the data more 'information' on the subject under enquiry, and therefore of leading to estimates of higher precision, and to tests of significance of greater sensitivity. And so it was in this atmosphere, borne in upon me that very often, when the most elaborate statistical refinements possible could increase the precision by only a few per cent, yet a different design involving little, or no additional experimental labour, might increase the precision two-fold, or five-fold or even more, and could often supply information in addition on relevant supplementary questions on which the original design was completely uninformative.

It was thus clear at an stage that there were quantitatively large technological gains to be obtained through the deliberate study of Experimental Design, and that these gains were to be harvested by making the plan of experimentation and observation logically coherent with the aims of the experiment, or, in other words with the kind of inference about the real world, which it was hoped that the experimental results would permit.

Fisher's thesis here is that when an experiment is improved, for example, by introducing better controls or other design considerations, the resulting data contain more Fisher-Information.

ILLUSTRATION. The research goal is to investigate the difference δ between two experimental treatments. There are two designs to choose from.

Experiment 1 — Arrange the field trials so that one treatment yields observations x_i ($i = 1, \ldots, n$) which are i.i.d. normal $N(\mu, \sigma^2)$ and the second treatment yields observations y_i ($i = 1, \ldots, n$) which are i.i.d. normal $N(\mu + \delta, \sigma^2)$. Suppose all three parameters are unknown, but that δ is the sole parameter of interest. This design might arise with random assignments to $2n$ plots in a given field. Half of the plots are randomly allocated to the first treatment and the remaining half are used for the second treatment.

Experiment 2 — Arrange the field trials so that $(x_i, y_i$ ($i = 1, \ldots, n$)) are n-matched pairs in the field, producing data from a bivariate normal population with unknown correlation ρ. Thus, as before, the data x_i ($i = 1, \ldots, n$) are i.i.d. normal $N(\mu, \sigma^2)$ and they y_i ($i = 1, \ldots, n$) are i.i.d. normal $N(\mu + \delta, \sigma^2)$, but the (x_i, y_i) pairs have correlation ρ. This design might be implemented by blocking the $2n$ plots into n pairs and randomly assigning one plot/block to each of the two test groups.

Which is the better experiment? Let us apply Fisher-Information to the resulting test estimators with which we shall conduct our inferences about δ. With design 1 (random assignment) there is a simple "Student's" t-test, based on the sample difference $(\overline{Y} - \overline{X})$, and a "pooled" variance estimate having $2(n - 1)$ degrees of freedom. In the second design (randomized blocks), there is a simple t-test based on the sequence of differences $z_i = y_i - x_i$ ($i = 1, \ldots, n$). Specifically, use a t-test with the sample average difference, $\overline{Z}$, and the associated variance estimate having $(n - 1)$ degrees of freedom.

Which design yields the test-statistic with the greater Fisher-Information about δ? The answer depends solely on the sample size n and the strength of the "matching," ρ. The table, below, lists the critical matching strengths (ρ) for which the second design has more Information about δ.

TABLE OF CRITICAL VALUES FOR ρ

n (number of paired observations)	ρ
2	0.167
3	0.160
4	0.143
5	0.127
10	0.0790
15	0.0569
26	0.0351
50	0.0188
250	0.0049
500	0.0020

Thus, the emphasis on match-pairs — a design concern — is justified by the fact that even with moderate samples sizes there is more Information about δ in the second design than in the first. Better to increase the precision by blocking (and use an estimate with n-1 degrees of freedom) rather than to double the degrees of freedom in a fully random allocation. Similar analysis indicates when Latin Squares, or other forms of restricted designs, yield estimates with greater Information about the parameters of interest. Information affords a measure of the efficacy in contemplated experimental controls.

2.2. The Problem of Randomization

Early in *The Design of Experiments* (§9—10) Fisher argues that randomization is a sine qua non of sound experimental practice. His now infamous pedagogical example, "The Lady Tasting Tea," offers three methodological lessons. We are to test the hypothesis that a certain lady cannot distinguish between tea made first by adding the milk as opposed to tea made by adding milk second.

The design calls for presenting her with 8 cups of tea with milk, 4 prepared each way, and counting how many the lady correctly identifies. According to Fisher's reasoning, by rigorously randomizing both the division of the cups (between the two treatments) and the order in which they are presented to the Lady, we achieve three goals:

(1) The random order of presentation is supposed to insure against the lady doing well on the test merely by anticipating the experimenter if, on an alternative design, the order of cups is decided by the experimenter. The randomization, then, is to prevent the experiment from becoming a game where the subject tries to outfox the experimenter.

(2) By randomizing the treatment allocation, the design is thought to insure against an unfortunate confounding of treatment with uncontrolled factors, which factors might be the actual cause of the lady's responses. For example, if the lady reacts to unobserved differences in the cups themselves, rather than to the tea mixtures, randomization establishes there is only a 1 in 70 chance the tea-milk combinations will align with cups so that she correctly identifies all 8.

(3) Randomizing the design, argues Fisher, provides a sound statistical basis for the resulting test of significance. That is, the randomization justifies the conclusion that, under the null hypothesis, there is a probability of 1/70 that all 8 cups are correctly identified, etc.

There is, however, a serious difficulty incorporating these arguments into the theory I am attributing to Fisher. That theory attempts to unify design with analysis, to use Information as the link between experimental design and statistical estimation. The problem centers on the role of ancillary data. (Recall, statistic T is ancillary for θ if it is probabilistically independent of the parameter of interest, $p(T \mid \theta) = p(T)$). In Fisher's theory of estimation, an ancillary statistic contains no relevant information about the parameter of interest. (The same conclusion follows according to Bayesian or Likelihood principles.) In numerous places Fisher takes pains to argue that, by conditioning on an ancillary statistic one creates estimates with greater Information.[8] Also, as in the illustration of the 2×2 table, conditioning on so-called "ancillary" data (the margin totals) may create an "exact" significance test for some composite null hypothesis.

The difficulty is simple to state: Randomization in design introduces ancillary data. That is, the outcomes of the randomization is ancillary to the hypothesis tests. Unfortunately, each of Fisher's three reasons for randomized design is based on probabilities which fail once the ancillary data of the randomization are given. That is, Fisher's arguments for randomization in design are valid using pretrial expectations; but they are unsupported post-trial, given the (ancillary) randomized outcome. There is no question that Fisher opposed randomization in analysis. He used the ancillarity principle to refute randomized statistical tests.[9] Nonetheless, his support for randomization in design remained undaunted. Even mild opposition from his longtime ally, Gosset ("Student," 1936), earned only (undeserved) scorn (Fisher, 1936a, b).[10] Among the several enigmas we owe to Fisher, coming to a proper understanding of the role randomization plays in sound experimental design is an ongoing activity. (See, for example, Rubin 1978.)

3. CONCLUSIONS

The position proposed here is that Fisher's contributions to the theory of experimental design are closely tied to his theory of estimation, with Information serving as the link. Estimation is concerned with data-reduction — where good estimators give summaries of evidence that preserve the relevant evidence as that is measured by Fisher-Information. Improvements in a design may be gauged by the increase in (Fisher-) Information of the resulting experimental data. Thus, there is a unified approach to statistical design and analysis. This approach serves, also, to explain the widely received view by statisticians that their role is not limited to post-trial consultation in data analysis. Statistics has its place in the planning of high quality experiments. In that sense, for an experimenter, it makes statistical sense to look before you leap!

The reconstruction I offer has difficulty, however, providing a rationale for the common methodological practice of randomized experimental design. The problem is that, in estimation Fisher's theory advocates conditioning on ancillary data. But randomization yields ancillary data. Then familiar (Fisherian) arguments fail when the same ancillary principle is applied to the outcome of the randomization. It is reassuring, at least, to know that the debates about randomization persist more than 55 years after Fisher's methodological innovation. (See, for example, the papers by I. Levi, D. Lindley, and P. Suppes in PSA-1982.)

Carnegie Mellon University

NOTES

[1] In SMfRW it is chapter 9, making up 40 pages out of 360. In DoE it is chapter 11, making up 35 out of 245 pages. And in SM&SI it is chapter 6, 35 of 180 pages.

[2] Fisher's earlier attempts to define consistency asymptotically, with increasing sample size, failed as they placed no restriction on how the estimator behaved in small samples. The version of F-consistency summarized here applies equally well to samples of arbitrary sizes, as Fisher notes (SM&SI, pp. 150—151).

[3] I find this assumption troubling. It seems plausible that the sample size is chosen in accord with the investigator's pretrial beliefs about the informativeness of the resulting data. But, except in rare circumstances, this judgment depends then upon the investigator's "prior" opinions about the parameter of interest. Hence, as a reader of the published data, I conclude that sample size is a function of the unknown parameter through the experimenter's "prior" for that parameter. That is, as a reader of the published data, I cannot take N to be ancillary without defaming the experimenter!

[4] See Savage (1976) for definitive rebuttals to these inaccuracies.

[5] See Fisher's statements in (SM&SI, §4.4) and (SMfRS, §21.02). The question is pursued by G. Barnard in three papers spanning the years 1946—1949.

[6] That is, given the sample sizes n_1 and n_2, the two quantities $(a/n_1 - c/n_2)$ and $(a/n_1 + c/n_2)$ are equivalent to the full data (a, b, c, d). Fisher's stipulation that the analysis be conducted for fixed lower margins amounts to a further data reduction to the test quantity, $(a/n_1 - c/n_2)$.

[7] The term $[T - (a + c)/N]^2 N \div T(1 - T)$ differs by a factor of N from C. R. Rao's measure of 2nd order efficiency. First order efficiency concerns the asymptotic ratio of information retained in an estimate. Second order efficiency concerns the difference between information retained and information available. Obviously, this limiting ratio may be 1 though the limiting difference is not 0. For example, minimum χ^2 estimation is not, though the m.l.e is 2nd order efficient in the 2×2 table. See Rao (1963) and Ghosh and Subramanyam (1974) for the key results.

[8] See especially Fisher's discussion of what he calls "The Problem of the Nile" (SM&SI §6.9).

[9] Fisher's opposition to a randomized solution of the Behrens-Fisher problem is found in SM&SI, §4.7. Additional discussion of this example may be found in Kadane & Seidenfeld (1990).

[10] In fairness, I believe Fisher allowed Gosset the last word in this exchange. Fisher left "Student's" (1937) final publication unanswered. That gesture, rare for Fisher, signifies his lasting respect for Gosset's contributions.

REFERENCES

Barnard, G. (1946), 'Significance Tests for 2×2 Tables', *Biometrika* **34**, 123—138.

Barnard, G. (1947), 'The Meaning of a Significance Level', *Biometrika* **34**, 179—182.

Barnard, G. (1949), 'Statistical Inference', *J. Royal Stat. Soc.* B **11**, 115—140.

Cramér, H. (1946), *Mathematical Methods of Statistics*, Princeton: Princeton Univ. Press.

Fisher, R. A. (1922), 'On the Mathematical Foundations of Theoretical Statistics', *Phil. Trans. Roy. Soc. London* A 22, 309—368.

Fisher, R. A. (1925a), 'Theory of Statistical Estimation', *Proc. Cambridge Phil. Soc.* 22, 700—725.

Fisher, R. A. (1925b), *Statistical Methods for Research Workers* (14th ed., 1973), New York: Hafner Press.

Fisher, R. A. (1934), 'Two New Properties of Mathematical Likelihood', *Proc. Roy. Soc. London* A 144, 285—307.

Fisher, R. A. (1935), *The Design of Experiments* (8th ed., 1966), New York: Hafner Press.

Fisher, R. A. (1936a), 'A Test of the Supposed Precision of Systematic Arrangements', *Annals of Eugenics* 7, 189—193.

Fisher, R. A. (1936b), 'The Half-Drill Strip System Agricultural Experiments', *Nature* 138, 1101.

Fisher, R. A. (1956), *Statistical Methods and Scientific Inference* (3rd ed., 1973), New York: Hafner Press.

Fisher, R. A. (1962), 'The Place of the Design of Experiments in the Logic of Scientific Inference', *Colloques Internationaux du Centre National de la Recherche Scientifique (Paris)* 110, 13—19.

Ghosh, J. K. and Subramanyam, K. (1974), 'Second Order Efficiency of Maximum Likelihood Estimators', *Sankhya* 36, 325—358.

Kadane, J. B. and Seidenfeld, T. (1990), 'Randomization in a Bayesian Perspective', *J. Stat. Planning and Inference* 25, 329—345.

Levi, I. (1983), 'Direct Inference and Randomization', in P. Asquith and T. Nickles (eds.), *PSA-1982*, vol. 2. Ann Arbor: Edwards Brothers, 447—463.

Lindley, D. (1983), 'The Role of Randomization in Inference', in P. Asquith and T. Nickles (eds.), *PSA-1982*, vol. 2. Ann Arbor: Edwards Brothers, 431—446.

Rao, C. R. (1963), 'Criteria of Estimation in Large Samples', *Sankhya* 25, 189—206.

Rubin, D. (1978), 'Bayesian Inference for Causal Effects: The Role of Randomization', *Ann. Stat.* 6, 34—58.

Savage, L. J. (1976), 'On Rereading R. A. Fisher', *Ann. Stat.* 4, 441—500.

"Student" (1936), 'Co-operation in Large-Scale Experiments', opening remarks by W. S. Gosset, Supplement to *Roy. Stat. Soc.* 3, 115.

"Student" (1937), 'Random and Balanced Arrangements', *Biometrika* 29, 363—379.

Suppes, P. (1983), 'Arguments for Randomizing', in P. Asquith and T. Nickles (eds.), *PSA-1982*, vol. 2. Ann Arbor: Edwards Brothers, 464—475.

JOHN MAYNARD SMITH

J. B. S. HALDANE*

The permeation of biology by mathematics is only beginning, but unless the history of science is an inadequate guide, it will continue, and the investigations here summarised represent the beginning of a new branch of applied mathematics.

These are the closing words of J. B. S. Haldane's *The Causes of Evolution*, published in 1932. Today, when one cannot discuss how some animal goes about finding its daily bread without basing the discussion on a mathematical model, it may seem strange that there was a time when ecology and evolution theory were free of mathematics. The investigations Haldane refers to are those carried out by R. A. Fisher, Sewall Wright, and himself into the genetic basis of evolutionary change. In fact, there had been an earlier phase in the mathematical treatment of evolution, by the biometrical school led by Karl Pearson; but this earlier approach was different, not merely in detail, but in its whole philosophy. For Pearson, the business of science was to provide a mathematical description of the world, but not to explain the behaviour of things in terms of some underlying mechanism. He went so far in this view as to deny the reality of atoms, because he thought it illegitimate to postulate the existence of unseen entities. It is not surprising that he also rejected genes. As a result, the work of the biometricians, although important in developing statistical techniques of value in the analysis of data, did little to illuminate the mechanisms of evolution. In contrast, Haldane and his contemporaries attempted to deduce, from the known principles of Mendelian genetics, how populations might be expected to evolve, and to compare their deductions with observation.

In doing this, they were applying to biology the method that had for a long time been the method of the physical sciences; hence, Haldane's reference to the history of science. In fact, the application of this method occurred more of less simultaneously in three different areas of biology: population genetics; ecology, in the work of Lotka and Volterra; and enzyme kinetics, a field to which Haldane also made a decisive early contribution. It is tempting to suggest that the early

37

Sahotra Sarkar (ed.), The Founders of Evolutionary Genetics, 37—51.
© 1992 *Kluwer Academic Publishers. Printed in the Netherlands.*

development of population genetics, which seems to have been carried out by three men who worked largely independently of one another, occurred when it did because there was a problem in evolutionary biology which could be solved only by the application of mathematics. This problem arose from the rediscovery of Mendel's laws in 1900. There soon developed two schools of thought. The Mendelians held that what was important in evolution was the origin of new types by mutation; the biometricians that the natural selection of continuously varying traits was what mattered, and that mutation gave rise only to monstrosities. No progress could be made until it was shown that the ideas of Darwin and Mendel were compatible. This was the main task facing Fisher, Wright, and Haldane. However, the simultaneous application of mathematics in ecology and enzymology suggests that there was something more general in the state of science in the twenties that favoured such developments.

Given the substantial degree of independence, at least in the early days, between the three men, it is interesting to look at the similarities and differences between them. The most striking difference between Haldane and Fisher (Wright, I think, is intermediate in this respect) lies in Haldane's interest in chemistry and biochemical genetics. He once remarked to me that he regarded chemical explanations as more fundamental than morphological ones. In 1932, there was little he could draw on in the way of biochemical genetics. At a fundamental level, he writes, 'Mendelian genetics suggests very strongly that even where variation is apparently continuous this appearance is deceptive. On any chemical theory of the nature of the gene, this must be so'; and later, 'to my mind it is probable that every gene produces a chemical effect, but we are very far from being able to prove this as yet.' At a detailed level, he was already interested in the genetics of flower colour because of the possibility of identifying the chemical effects of genes. In his lectures and writings, he frequently refers to the importance of 'inborn errors of metabolism' as a guide to the nature of genes. But perhaps the most characteristic example of his interest in the chemistry of the gene lies in his pre-vision of the famous experiment by which, in 1958, Meselson and Stahl demonstrated the semi-conservative nature of gene replication. Writing in 1941, in *New Paths in Genetics*, he asked how genes replicate.

We cannot imagine the gene swelling till it divides like an overgrown drop of water . . .

the gene is spread out in a flat layer, and acts as a model, another gene forming on top of it from pre-existing material such as amino acids. This is a process similar to crystallization. . . . How can one distinguish between model and copy? Perhaps you could use heavy nitrogen atoms in the food supplied to your cell, hoping that the 'copy' genes would contain it while the models did not.

In 1932, a chemical theory of the gene was still 20 years in the future. Yet Haldane's wish to have a materialist, empirically based theory of evolution is very apparent in *The Causes of Evolution*. A surprisingly large part of the book is concerned with what was then known about genetics. In discussing the nature of species, he attempts to show that all the kinds of difference, genic and chromosomal, that are found between species can also be found between members of the same species. The only photograph in the book is of the artificial allotetraploid 'species' *Primula kewensis* and its parents, *Primula floribunda* and *Primula verticellata*.

I will discuss later what I believe to be the explanation for Haldane's fascination with biochemistry. Clearly, it is not enough to say that it arose because, during the period 1923—33 when he did his early work in population genetics, he was a member of the biochemistry department at Cambridge. That would not explain how a man whose undergraduate training was in mathematics and the classics came to be a biochemist. But first I must say something about the population genetics in *The Causes of Evolution*. I will do this by describing what Haldane had to say about two topics that are of contemporary interest: speciation and the evolution of altruism.

Rereading *The Causes of Evolution* recently, I was surprised to find that Haldane's views on speciation at that time were much closer to those of Wright than were the views he later expressed in the undergraduate lectures that I attended at University College London in 1950. He discusses the two-locus case, in which the double dominant AB and the double recessive $aabb$ are fitter than the types Abb or aaB. In 1931, he published an analysis of this case in which he presents the results as trajectories in a two-dimensional state space, showing the unstable watershed and the attractors at AB and $aabb$. (He remarks that he has been unable to find explicit equations for the trajectories in the three- and four-locus cases.) In the book, he writes: 'The change from one stable equilibrium to the other may take place as the result of the isolation of a small unrepresentative group of the population, or a temporary change in the environment which alters the relative viability of the different types.' He later considers the possibility of a peak shift in a large population with a high frequency of selfing or inbreeding. He continues:

This case seems to me to be very important, because it is probably the basis of progressive evolution of many organs and functions of higher animals, and of the breakup of one species into several. For an evolutionary progress to take place in a highly specialised organ such as the human eye or hand a number of changes must take place simultaneously. Thus if the eye is unusually long from back to front we get shortsightedness, which would not, however, occur if there were a simultaneous decrease in the curvature of the cornea or lens, which would correct the focus. . . . Actually a serious improvement of the eye would involve a simultaneous change in many of its specifications.

These ideas are close to Wright's 'shifting balance' theory, although Haldane does not discuss the evolutionary potential of a population divided into many partially isolated demes. I think Haldane reached his conclusions independently of Wright. In the appendix to *The Causes of Evolution*, he gives an account of Wright's 1931 paper in *Genetics*, but says that his own book (based on lectures given in 1931) was written before Wright's paper appeared. The convergence is therefore interesting, particularly the parallel between Haldene's use of trajectories in state space and Wright's concept of an adaptive landscape.

This tempts me to a digression on the use of visual imagery. Fisher had remarkable intuition, but it seems to have been almost entirely non-visual. It may be for this reason that many biologists find him hard to follow. In contrast, Wright's adaptive landscape is perhaps the most influential image in biology. Haldane liked to use graphical methods to solve problems, although he was also liable to plunge into long algebraic calculations. ('All that algebra that Jack seems to find necessary,' to quote Fisher.) One example of Haldane's use of graphs is worth describing. In 1950, finding that, perhaps for the first time, he had a mathematically literate undergraduate in his class, he gave a special series of lectures on mathematical ecology and population genetics. Figure 1 shows his treatment of population number when generations are separate. He pointed out that, as the graph of X_{n+1} against X_n changes, the population first approaches its equilibrium value without oscillations, then shows damped oscillations, then sustained oscillations of period 2, and then oscillations of period 4. After that, he said, 'Something odd seems to happen.' Unfortunately, his only

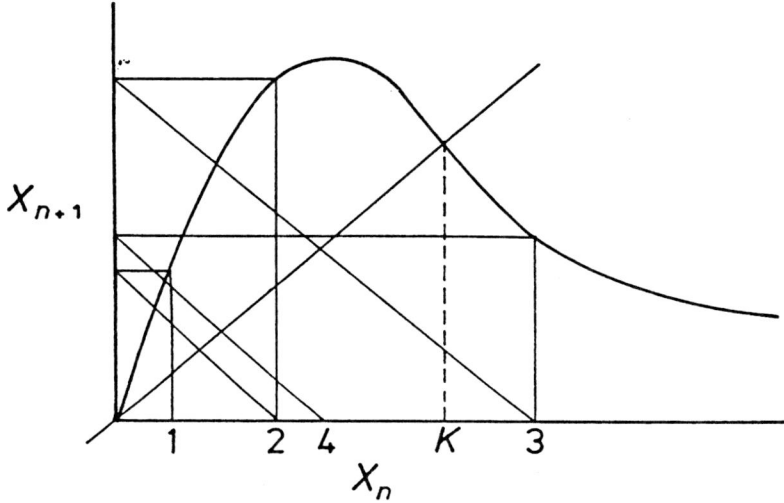

Fig. 1. A graphical method of calculating successive values of the size of an annual population. The bold curve gives the number X_{n+1} in year $n + 1$ as a function of X_n, the number in the year n. The numbers 1, 2, 3, and 4 represent successive values of X_n. K is the equilibrium value of X_n, or 'carrying capacity'.

calculating aids were pencil and paper; he must have been one of the last people to use the old method of long division. When I tried to repeat his calculations, I used a slide-rule, a method even less likely to lead to a correct conclusion. Had either of us used a computer, we might have written the paper on chaos in ecology that Robert May was to write in 1976. The story echoes the one told above of Haldane's proposal to label genes with heavy nitrogen: the right idea, but inadequate technology.

Sadly, I have lost my notes on these lectures, but some random memories may be of interest. He discussed animal life tables, suggesting that mammals may senesce in the wild but birds do not. However, in ecology his central concern was the natural regulation of animal numbers. In general, he sided with Lack and Nicholson rather than Andrewartha and Birch: that is, he was seeking density-dependent mechanisms that limit populations. He was interested in the possibility that parasites, rather than food, might regulate animal numbers, and spent some time on the beautiful host-parasitoid model of Nicholson and Bailey, and the data that might support it. Surprisingly, I remember less of his lectures on population genetics, although I remember him returning very excited after a day in Oxford with Cain and Sheppard. I also recall my own excitement when he described Malecot's elegant treatment of the approach to homozygosity in an inbred line. I also remember deciding, after a lecture on McClintock's jumping genes, that I would avoid the topic if it came up in finals. The lectures were mainly remarkable, not for any specific topic, but as an illustration of how to make models in biology.

Returning to Haldane's population genetics, in *The Causes of Evolution* he discusses the evolution of 'socially valuable but individually disadvantageous characters'. In the appendix he presents a formal model, which is essentially a 'group selection' model similar to my 'haystack' model but worked out in less detail. He concludes that the spread of an 'altruistic' gene requires that group size be small and endogamy fairly strict, and ends, 'I find it difficult to suppose that many genes for absolute altruism are common in man.' By 'absolute altruism' he seems to mean altruism not directed specifically towards relatives, except in so far as any member of one's own endogamous group will be a relative. Thus, he is clear, as everyone must be, that altruism towards direct descendants can be favoured by selection, even in a large unstructured population; but he also extends this argument to other relatives. For example, he writes, 'For in so far as it makes for the survival of one's descendants *and near relations*, altruistic behaviour is a kind of Darwinian fitness, and may be expected to spread as a result of natural selection' (my emphasis).

Having appreciated the ineffectiveness of group selection and come so close to formulating the theory of kin selection, why did he not proceed to develop the ideas later formulated by Hamilton, according to which co-operation in animal societies has evolved because of the genetic relationship of their members? Part of the answer must be that the problem of the evolution of social behaviour was not his central concern, which was to demonstrate the compatibility of Mendelian genetics with Darwinian natural selection. But this cannot be the whole answer; like Fisher and Wright, he was fascinated by the problem of altruism. The question becomes still more puzzling because, in 1955, he showed that he understood the arithmetic of Hamilton's rule, $rb > c$, at least in cases in which the gene for altruism is rare.

One cannot answer this question with confidence; but I think that the failure arose because he was asking the wrong question, at least as far as non-human animals are concerned. It seemed clear that, in our own species, individuals do things that lower their own fitness and benefit other, wholly unrelated members of the population. Haldane, for example, mentions Christian saints and winners of the Victoria Cross. The question therefore arose as to whether such 'absolute' altruism could evolve as a genetically determined trait. The answer seems to be that it cannot, except in species with a very unusual breeding structure. The conclusion, then, is that cultural rather than genetic explanations are needed for absolute altruism. The evolution of altruistic behaviour between relatives then seems to be largely irrelevant. It is interesting that, in discussing social insects, Haldane points out that, given that there are sterile castes, it is understandable that their members should display extreme altruism; but he does not ask how such sterile castes arose in the first place. Instead of attempting to explain the evolution of the more extreme forms of human self-sacrifice, it might have been more fruitful to ask whether altruism between relatives can account for social behaviour in animals. Most biologists are now persuaded that relatedness is at least highly relevant; but we would not have reached that conclusion if our central concern had been to explain the behaviour of Christian martyrs.

This leads me to make a comment on the significance of natural history. However great his passion for general ideas, the roots of Darwin's thinking lay in his knowledge of natural history. The same is true of Wallace. It is also true, I believe, of the architects of the 'modern synthesis', in particular Mayr and Stebbins, and of men like Tinbergen, Lack, and Hamilton, whose work has led to our current understanding of the evolution of behaviour. I have the strong impression, reading some of the criticisms of sociobiology, that the critics genuinely cannot conceive that workers in the subject are interested primarily in animals. Surely, there must be some hidden agenda concerned with human affairs. Of course, in some cases, most notably that of E. O. Wilson, there is such an agenda, although it is not a hidden one. All of us, no doubt, have political and ideological commitments that influence our science. But it is worth emphasizing that, for all the people mentioned above (including, of course, Wilson), a familiarity with and love of natural history has been both an important motivation and a source of questions and hypotheses. Surprisingly, I do not think that this was true of the founding fathers of population genetics. Haldane, certainly, was not primarily a naturalist, although he was well aware that the testing of ideas about natural selection required field studies, and was therefore enthusiastic about the work of people like Lack, Tinbergen, Cain, and Sheppard. But the sources from which he drew his inspiration were the physical sciences, literature, and mythology, not natural history. I think it was for this reason that he asked the wrong, or at least the less fruitful, question about the evolution of altruism.

Although his discussion of the evolution of altruism must, I think, be classified as a near miss, he was from the beginning clear about 'the good of the species'. A paragraph from *The Causes of Evolution* is worth quoting in full:

But I must leave this fascinating topic [the role of chance in speciation] to discuss a fallacy which is, I think, latent in most Darwinian arguments, and which has been responsible for a good deal of the poisonous nonsense which has been written on ethics in Darwin's name, especially in Germany before the war and in America and England since. The fallacy is that natural selection will always make an organism fitter in its struggle with the environment. This is clearly true when we consider the members of a rare and scattered species. It is only engaged in competing with other species, and in defending itself against inorganic nature. But as soon as a species becomes fairly dense matters are entirely different. Its members inevitably begin to compete with one another. I am not thinking only of the active and often conscious competition between higher animals, but also of the struggle for mere space which goes on between neighbouring plants of closely packed associations. And the results may be biologically advantageous for the individual, but ultimately disastrous for the species. The geological record is full of cases where the development of enormous horns and spines (sometimes in the male sex only) has been the prelude to extinction. It seems probable that in some of these cases the species literally sank under the weight of its own armaments. Again, while modern research tends to show that sexual selection in birds is rather less important in making bright colour and structures such as the peacock's tail advantageous in male birds than Darwin supposed, there is still a good deal of evidence that it has certain selective value in securing mates. And none will contend that (except in so far as it has induced Hindus to regard him as sacred and Europeans as a suitable pet) the peacock's rather cumbrous tail has been of any advantage to him in the struggle with the environment.

I first learnt this point when I wrote an undergraduate essay for Haldane, in which, unwittingly, I explained the evolution of some trait in terms of species benefit. He wrote in the margin, 'Pangloss' theorem'. I knew that the theorem must be: 'All is for the best in the best of all possible worlds', but it took me some time to see what he was getting at. I think the effort was worth while. 'Pangloss' theorem' has sometimes been interpreted as implying that all individuals are perfectly adapted to their environments and free of historical constraints. I think that Haldane saw it as supporting the view that natural selection will always benefit the species. It is important to distinguish between these two fallacies. If I am right, Haldane invented "Pangloss' theorem" as a way of criticizing "good of the species" arguments. More recently, it has been used to criticize optimisation theory in biology, on the grounds that historical and genetic constraints will prevent a species reaching an optimum. This is a respectable argument (although I think it has been over-played), but it is not the one Haldane was concerned with.

I have discussed peak shifts and the evolution of altruism at some length because of their contemporary interest; but they would have seemed relatively minor matters to Haldane. The main topics of *The Causes of Evolution* are the nature of species differences, the possibility that natural selection can produce complex characters, the causes of genetic variability and polymorphic equilibria, and the relative importance of natural selection and mutation. In later years, his most significant contributions were probably to human genetics and to the concept of the 'cost of selection', which is the origin of the theory of genetic loads.

Why did Haldane become an evolutionist and a biochemist? I do not think that an interest in eugenics played a major part, as it seems to have done with Fisher. Haldane's view of eugenics, expressed in his essay 'The Inequality of Man', was that men were not created equal, and that it was therefore desirable to organize society so that people of varied constitutions could be useful and contented in it. He was highly critical of proponents of sterilization on eugenic grounds, whom he referred to as the 'Off with your cock brigade'. He had little sympathy with positive eugenics — that is, attempts to produce outstandingly talented individuals, by artificial selection by selected donors as suggested by H. J. Muller, or in other ways. He was interested in negative eugenics — the reduction in the frequency of births of individuals with specific genetic disabilities — but he was opposed to sterilisation, and reserved about selective abortion, because of the potential abuses of such measures. To reduce the frequency of individuals homozygous for deleterious recessives, all we need to do is to reduce the frequency of marriage between people carrying the same deleterious genes. He once suggested, I think in conversation, that, as we become able to diagnose the presence of more such genes in heterozygotes, people should be encouraged to wear jewelled pins, with different types of stone symbolising different genes. It would then be natural and easy to avoid a close relationship with someone wearing the same type of pin. He was of course aware that such a measure, if successful, would reduce the frequency of homozygotes at the expense of relaxing selection against deleterious recessives, and so allowing their frequency to rise, but he thought that the time scale is such that we will have discovered more direct ways of altering genes before the frequency increase is detectable.

I think that a clue to his interests may lie in a striking feature of his writing and conversation. This is the ubiquity of references to religion and of quotations from religious texts. These references are not only to Christianity, but to classical mythology and, after he settled in India, to Hindu and Buddhist writings. This is not an attempt to use religious arguments to support his case: Haldane was not a believer and, at least as far as Christianity was concerned, he rather enjoyed blasphemy. Neither was it, in any simple sense, an attempt to settle scores with a religion he had abandoned. I mention this possibility because such a motive has been important in my own case; having been raised in a rather literal version of Christianity, my interest in Darwin was first aroused because of the apparent contradiction between the theory of evolution and the Christian faith. Other evolutionary biologists have had a similar experience: E. O. Wilson is, I believe, an example, and I was amused to discover how many of my co-workers are lapsed Catholics. But Haldane seems never to have been a religious believer.

I think that Haldane's repeated references to religion reveal two fundamental aspects of his thought. One was an interest in philosophical questions, to which I will return in a moment. The other was that myths are the most effective means of moral instruction, whether or not one believes in their literal truth. Repeatedly, in conversation and writing, he would make a moral point by quoting the experience of a mythical hero. For example, in one of his last essays, on non-violence, he wrote:

My attitude to animals is more like that of Yudhisthira. He had killed and eaten plenty of deer. But when he was asked to enter *Svárga*, leaving the dog which had been his companion on his last pilgrimage to die on the mountains, he found, perhaps to his surprise, that this was something which, as a Kshattriya, he could not do.

Life with Haldane required, among other skills, that one brush up one's mythology.

But it is his philosophical interests that are more relevant to his science. They originate, I think, in interactions with his father, the physiologist J. S. Haldane. J. B. S. was encouraged to act as his father's assistant, and, to the end of his life, the only experimental techniques which he ever mastered were those concerned with the handling of gases, which he learnt as a boy. Philosophically, J. S. Haldane was an idealist; in particular, he thought that there was something special about living matter which was incomprehensible in purely physical and chemical terms. J. B. S. loved his father, but he spent his life trying to prove him wrong.

In 1920, there were three places where the essential mystery of life might be supposed to reside: in the chemistry of protoplasm, in the evolution of new forms, and in the origin of consciousness. When he died, J. B. S. could reasonably hold that there were few hiding-places left for mysterious forces in biochemistry or evolution; but he was still thinking about consciousness.

The last chapter of *The Causes of Evolution* is essentially philosophical. Today, evolutionary biologists are nervous of such discussion, particularly since Griffin invented the term 'philosopause'; but Haldane had no such inhibitions. He states his basic position as follows:

Particularly hostile to true scientific progress are the extremer forms of the doctrine of emergence. According to these, a material system of a certain degree of complexity suddenly exhibits qualitatively new properties such as life or mind, which cannot be explained by those of the constituents of the system. There is clearly an element of truth in this view. We can only discern a little mind in a dog, and at present none in an oyster or an oak. Nevertheless science is committed to the attempt to unify human experience by explaining the complex in terms of the simple. This may be a vain endeavour, but I do not at present see any evidence of its vanity.

In thinking about the evolution of consciousness, Haldane relied on an analogy with atomic physics (remember that the date is 1932). The hydrogen atom has complex modes of vibration, reflected in its spectra. If the proton and the electron are spherical electric charges, this behaviour is incomprehensible; the hydrogen spectra would then be emergent properties, incomprehensible in terms of the properties of its components. But it turns out that the electron itself has undulatory properties; and electron beam forms diffraction patterns. Once the undulatory nature of the electron is appreciated, the properties of the hydrogen atom are no longer incomprehensible. By analogy, Haldane suggested that the consciousness of man must ultimately depend on mind-like properties of atoms. This is an idea to which he returned several times.

The passage quoted expresses a view that today would be called reductionist. It was, of course, written before he joined the communist party and became a Marxist, but it does raise the question of how far Marxism affected his science. Very little, in my opinion. One can, I think, detect a change in his attitude towards human genetics between "The Inequality of Man" (1932) and "Heredity and Politics" (1938). In the former he argues that all men are not created equal, and enquires how society might be structured so that different kinds of people can live happy and useful lives. In the later book he emphasizes the complexity of the interactions between nature and nurture, and the dangers of eugenic legislation. However, I think it was the rise of Hitler, and of Nazi race theories, that caused both his changed views on eugenics, and his joining the communist party: I do not think there was any direct effect of Marxism on his views on eugenics.

In general, his method of doing science seems to have been little altered by Marxism. One can, I think, argue that when he was influenced by Marxism, he was misled by it. I have already discussed his failure to develop the idea of kin selection as a cause of the evolution of animal societies, and have suggested that the reason was that his interest at the time lay primarily in the origins of human altruism, and not in the natural history of ants. This may have been compounded by the more general objection felt by Marxists to biological explanations in the human sciences. I can offer two other, rather trivial, examples of Haldane being misled. In an essay, "Beyond Darwinism", published in the Daily Worker, he praises Elton for expressing a group-selectionist explanation of population regulation (similar to that later proposed by Wynne-Edwards), and adds "Elton is not, so far as I know, a Marxist. But I am sure that Marx would have approved his dialectical thinking". I find this particularly ironic, because I first learnt to distrust group-selectionist thinking from Haldane, some ten years after the above essay was written. A second example is equally curious. I quoted earlier his extraordinary prevision of the Meselson-Stahl experiment demonstrating semi-conservative replication. But in the same lecture he goes on to point out that the exchange of material between the gene and the surrounding fluid might make such an experiment inconclusive, and concludes "I think that throughout genetics an attempt to impose mechanistic interpretations such as the model and copy theory will break down . . . . However, a refutation of mechanism is not a refutation of materialism".

On these occasions, Haldane's faith in dialectics seems to have been excessive, but no great harm was done. More important issues, however, are raised by the Lysenko affair. Before discussing Haldane's attitude, I must digress to discuss my own. I also joined the communist party in the late thirties, mainly because it seemed the most consistent opponent of the Nazis. When, after the war, I decided to leave engineering and study biology, I went to UCL because Haldane was there. During the period 1945-1955, I became increasingly disillusioned, and finally left the party in 1956. The Lysenko affair was an important cause of my disillusion. People sometimes find this puzzling: faced with the gulags, why should Lysenko matter? The point, of course, is that I dismissed what I read about the gulags as anti-soviet propoganda. But I knew about genetics, and the account of the meeting of the Lenin Academy of Agricultural Sciences in 1948 was published by the Russians themselves. Lysenkoism was important to me because, if the central committee of the Soviet communist party could be wrong about that, what else could they be wrong about? It was the crack in the dyke.

How, then, did I see Lysenkoism? The idea of the inheritance of acquired characters did not seem to me obviously false: indeed, I was prejudiced in its favour. There is something deeply undialectical about a gene that influences development, but is itself unaffected. I therefore do not think that those Marxist philosophers who supported Lysenko were merely jumping on a bandwagon, although doubtless some were. If they sincerely believed that Marxism was a good guide to scientific practice — and I certainly thought that in 1948 — then they were right to support Lysenko. There was then, and is now, some evidence for the hereditary transmission of acquired traits. So why did I find Lysenkoism disillusioning? There were several reasons. First, Lysenko and his supporters rejected Mendelism out of hand. Now Mendelism, like Newtonian mechanics, may be a special case, but it is not merely wrong. Indeed, it was obvious from the debate that Lysenko's supporters did not understand the views they were criticising. More important, I was horrified that the central committee of the Soviet party felt itself able to decide that Lysenko was right, and that, as a result, his opponents were forced to make humiliating apologies, and many lost their jobs. Finally, I was disgusted by the way in which some British communists, who knew little of genetics, felt free to tell me what I ought to think.

I have digressed to describe my own thoughts because I am aware of the danger, on this and on other matters, of assuming that, because I thought something, Haldane must have thought so too. In fact, I never discussed these matters with him in any depth, although I heard him speaking on the subject at inner-party meetings, and was present on a number of more public occasions when the subject came up. I am reasonably confident of the following. First, he took Lysenko's scientific ideas seriously, and tried to find out as much as he could about the experimental evidence. It may be that one reason for this is the point I made above, that Marxist dialectic would prejudice one in favour of Lysenko, but I do not remember that he used this argument. An argument I do remember him using was that we should take seriously any alternative to our favourite orthodoxy. Within the communist party he argued fiercely that the party should not commit itself to the view that Lysenko was right, and that it was wrong that reputable Russian scientists were being dismissed. I guess, but do not know, that for Haldane as well as for me the Lysenko affair was a catalyst that led to wider disagreements with the party. I do know that, when I told him I had left the party, he seemed delighted.

To have worked with Haldane for 10 years is an experience that is not easy to describe. I went to University College London as a student because I knew that he taught there. I already shared many of his tastes and prejudices: a materialist philosophy, a liking for simple mathematical models, and a dislike of the British upper class. There were also differences: my approach to biology is informed by a love of natural history and, most emphatically, not by an understanding of chemistry. Temperamentally, I like a quiet life while Haldane did not. But the similarities of outlook, coupled with the differences in age and intellect, mean that I often do not know whether an idea is really my own or borrowed from Haldane. Conversely, when asked about Haldane's view on some matter, I tend to assume that he must have thought as I do. Rereading *The Causes of Evolution* and his early essays has done little to clear up the confusion. I have merely found myself falling again under the influence of an extraordinary mind.

ACKNOWLEDGEMENT

* This is an expanded version of the article: Maynard Smith, J., "J. B. S.

Haldane", *Oxford Surveys in Evolutionary Biology*, 1986, by permission of Oxford University Press.

University of Sussex

REFERENCE

May, R. M. (1976), 'Simple Mathematical Models with Very Complicated Dynamics', *Nature* **261**, 459—67.

SAHOTRA SARKAR

HALDANE AS BIOCHEMIST:
THE CAMBRIDGE DECADE, 1923—1932[1]

1

From 1923 until 1932 J. B. S. Haldane served as Reader in Biochemistry at Cambridge under Sir Frederick Gowland Hopkins. This
decade was critical to Haldane's intellectual development. During it,
along with Fisher and Wright, he laid down the mathematical foundations of evolutionary genetics, the contribution that remains his most
important claim to fame (Wright 1968). The first part of his seminal
series of papers, "A Mathematical Theory of Natural and Artificial
Selection," appeared in the *Transactions of the Cambridge Philosophical Society* in 1924 (Haldane 1924). Only the tenth and last part
(Haldane 1934) post-dated his tenure at Cambridge. Meanwhile, in
1932, he had published *The Causes of Evolution* (1932), the main text
being based on a set of lectures given at Prifysgol Cymru (Wales) in
1931. Fisher's *Genetical Theory of Natural Selection* had appeared in
1930 (Fisher 1930); Wright's long paper, "Evolution in Mendelian
Populations," in 1931 (Wright 1931). The mathematical foundations
for evolutionary genetics had finally been laid down: not only had
Mendelian genetics been shown to be consistent with Darwinian evolution by natural selection but had, in fact, been shown to provide the
principle of inheritance which explained much of the phenomena of
evolution. Haldane has routinely received full credit for his role in these
developments: along with Wright and Fisher he is regarded as one of
the three founders of mathematical population genetics (see, e.g.,
Provine (1971)).

What is less known, however, is that during this decade at Cambridge, Haldane's work on the mathematical foundations of evolutionary genetics constituted but a mere fragment of his scientific output.
His interests were significantly spread over at least four other topics.
First, with various collaborators, he continued experimental work in
physiology in the tradition learnt from his father, J. S. Haldane. *Second*,
he made some critical contributions to the study of the kinetics of
enzymes. *Third*, he initiated, through a collaboration between the

53

Sahotra Sarkar (ed.), The Founders of Evolutionary Genetics, 53—81.
© 1992 *Kluwer Academic Publishers. Printed in the Netherlands.*

Biochemistry Laboratory at Cambridge and the John Innes Horticultural Institute at Merton, an important research program in chemical
genetics. *Fourth*, he continued with other genetical studies, including
studies of linkage and the relation between genetics and development.

The purpose of this paper is to describe Haldane's work in biochemistry — on enzyme kinetics and biochemical genetics — in some
detail and to put it in its proper intellectual context. There are two
general reasons why this is important. *First*, this work is quite important
and has usually either not received any attention, has been denigrated,
or significantly misrepresented.[2] For instance, Haldane's biographer,
R. W. Clark, thought that what Haldane had achieved was a demonstration of the fact that enzymes obeyed the laws of thermodynamics,
as if that was a controversial point that needed to be settled (1969,
68; 1972, 22)! The historian, R. C. Olby, though certainly aware
of Haldane's work, has assessed that there was nothing positive or
original in Haldane's contributions to biochemical genetics (1974, 136).
Even G. Werskey, who seems to have been more cognizant of the
relative worth Haldane's work in biochemistry, gives a basically negative assessment (1972, 82—83). The considerations below, besides
providing an account of this work, will implicitly argue against such
assessments.

Second, and more important, Haldane's work under Hopkins led to
a fundamental shift in his philosophical outlook in a way that had a
critical bearing on all his subsequent work. When Hopkins offered
Haldane the job at Cambridge, Haldane accepted but observed: "It is
only fair to say that most of the ideas which I have on physiology are
really [those of J. S. Haldane]. I tend to think of physiological questions
primarily in terms of 'milieu intérieur' rather than metabolism, thanks to
him and Claude Bernard, and this enables one to see problems which
from the point of view of method, are much simpler than many
metabolic ones (1922)." By the time Haldane left Cambridge, he had
initiated a program of research to elucidate the genetic control of
metabolic processes leading to pigment appearance in plants. The
biochemical point of view pervaded *The Causes of Evolution* (1932).
He had also attempted to give a physical account of the origin of life
(Haldane 1929). Though not yet a materialist like Hopkins — that
conversion would come a few years later as he drifted to the Communist Party — the search for material biochemical explanations had
already become central to Haldane's concerns.[3] The search for func-

tional organization and explanation that so dominated the work of J. S. Haldane had all but disappeared, as had any serious interest in physiology. The critical factor in these shifts was the influence of Hopkins during the years spent under him, doing biochemistry at Cambridge.

2

At first sight, Hopkins' choice of Haldane as the first Sir William Dunn Reader in Biochemistry does seem somewhat paradoxical as Werskey (1972, 83) has observed. Hopkins had come to Cambridge in 1898 at the invitation of Michael Foster, then Professor of Physiology.[4] He was supposed to teach and conduct research on what was then known as "chemical physiology," which involved teaching both chemistry and anatomy.[5] Biochemistry was emerging as a self-consciously separate discipline only at this time. In fact, it seems to have emerged quite suddenly during 1901—1905 though the term had been used by Hoppe-Seyler as early as 1877.[6] Much of Hopkins' organizational efforts during the first two decades of this century was directed at establishing biochemistry as a separate discipline independent from physiology. It was not easy going. At Cambridge, he only obtained a praelectorship from Trinity College in 1910 and when, in 1912, he was finally given a chair of biochemistry, it was unpaid.[7] It was only through the intervention of his friend, Walter Fletcher, who had emerged as an influential statesman of British science in the period following World War I, that the trustees of the estate of Sir William Dunn finally agreed in 1920 to fund an institute for Hopkins' research. Hopkins was able to draw the interest in the endowment starting with 1921; a new laboratory was built and officially opened on May 9, 1924. The grant included a Readership which Hopkins had offered to Haldane in 1922 and Haldane had accepted.

Haldane, at that time, was a Fellow at New College, Oxford, a position that he had accepted just before the end of World War I, and which had begun in October, 1919. He had been educated at Eton, which he seems not to have enjoyed, and then at New College, where he took Firsts in Mathematics and Classics. He had no formal training in biology except for his attendance at a course in vertebrate anatomy at Oxford (Haldane ca. 1942). However, his privileged background introduced him to biology at an early age. His father, John Scott

Haldane taught physiology at Oxford and laid down, over a period of serveral decades, the principles of respiratory physiology. The younger Haldane had assisted his father with experiments from childhood and had provided the mathematical analysis in a very important paper establishing the laws of the combination of haemoglobin with oxygen and carbon monoxide (Douglas *et al.* 1912).[8] He had also been fascinated with genetics since 1901 when, at the age of eight, his father had taken him to a lecture at the Oxford University Junior Scientific Club, to hear A. D. Darbishire lecture on the rediscovery of Mendel's laws. By 1912 he had observed that some data reported by Darbishire (1904) revealed the existence of linkage in mice, a point that Darbishire had not noticed. Advised by Punnett (1912) to continue with his own experiments, he began research in collaboration with a friend, A. D. Sprunt, and his sister, N. M. Haldane. These resulted in a report published in the *Journal of Genetics* in 1915 (Haldane *et al.* 1915), one of the first reports of linkage in higher animals. World War I, in which Haldane served with characteristic bravado and distinction, interfered with these researches but he returned to them in New College. While there he published six papers in genetics and five in physiology between 1919 and 1922. To biochemistry he seems not to have paid much attention. Though he spent the summer of 1921 as a biochemist at the Edinburgh Royal Infirmary, a delay in the completion of the construction of laboratory facilities had prevented him from carrying out any research (Haldane *ca.* 1942).

Yet, it was to Haldane that Hopkins, who had barely won a long struggle to establish biochemistry as a separate discipline from physiology at Cambridge, offered the endowed Readership which made Haldane, in effect, second-in-charge of his department. However, the choice is not as paradoxical as it initially seems mainly because the physiology of that period included considerable research that later formed part of biochemistry. In fact, it seems fair to say that the difficulties experienced by biochemistry, as it emerged as an independent science, were to a large extent due to institutional and professional difficulties experienced by the "new" biochemists, not because of any major conceptual or experimental shifts that physiology could not have accommodated.[9] Hopkins wanted to pursue biochemistry as a fundamental discipline independent of physiology partly because of desire to orient it away from pathology, to which physiology then paid considerable attention, and also in order to concentrate on biochemistry alone, that is, to ignore other aspects of physiology.[10]

That at the level of the direction of their research Hopkins and Haldane were not far enough apart to generate the paradox mentioned above is best illustrated by a detailed consideration of Hopkins' famous address, 'The Dynamic Side of Biochemistry,' delivered at the British Association for the Advancement of Science in 1913 (Hopkins 1913). It is here that Hopkins lays out, in full, his agenda for biochemistry emphasizing that the need for trained biochemists would only increase in the immediate future. There are four aspects of this address that are particularly relevant here. The first two concern Hopkins' conception of biochemistry. *First*, Hopkins emphasized that biochemistry was concerned "not alone with the separation and identification of products from the animal that out present studies deal; but with their reactions in the body; with the dynamic side of biochemical phenomena (1913, 137)." Thus was introduced the notion of "dynamic biochemistry" that became almost a catchword during the next decades (Kohler 1973, 183).

Second, according to Hopkins, "in the study of the intermediate processes of metabolism we have to deal, not with complex substances which elude ordinary chemical methods, but with simple substances undergoing comprehensible reactions (1913, 137)." The point was critical: Hopkins was appealing to organic chemists, trained in exact methods, to shift at least part of their attention to biochemistry. As he observed later in the address: "One reason which has led the organic chemist to avert his mind from the problems of Biochemistry is the obsession that the really significant happenings in the animal body are concerned in the main with substances of such high molecular weight and consequent vagueness of molecular structure as to make their reactions impossible to study by his available and accurate methods (1913, 144)." Instead, according to Hopkins, the important molecules were small and, therefore, simple enough for such study. He gave some putative examples.

Third, Hopkins was particularly sensitive to the quite common charge that the experimental methods of biochemists were distrusted by the chemists on grounds of sloppiness (1913, 138—139). Though he did not outright admit to this charge and, in fact, suggested that remarkable progress had been made since his youth regarding the exactness of the experimental methods of biochemistry, he underscored the impor-

tance of such exactness. In fact, as should be evident from what has
been noted above, it was in order to emphasize the possibility of such
exactness that Hopkins argued that the relevant molecules were sufficiently simple for exact study.

Fourth, Hopkins argued for interdisciplinary training for future
workers in biochemistry (1913, 145). If their background was in
organic chemistry, then they needed biological training in order for
them to have a proper orientation towards biological problems: "it
should be clearly understood," Hopkins argued, "that the progress
made in [past biochemical research] could only have come through the
work and thought of those who combined with chemical knowledge
trained instinct and feeling for biological possibilities (137, 145)." After
all, the problems that biochemistry was to solve were set, ultimately, by
a biological context.

About one year before this address, there had appeared in the
Journal of Physiology a critically important paper by C. G. Douglas,
J. S. Haldane and J. B. S. Haldane entitled "The Laws of Combination
of Haemoglobin with Carbon Monoxide and Oxygen" (Douglas *et al.*
1912). The experiments reported were due to the first two authors, the
mathematical analysis entirely due to J. B. S. Haldane. The experimental results consisted of a series of dissociation curves for the
reaction between carbon monoxide and hemoglobin in the blood of
Douglas, J. S. Haldane and two mice, in various controlled laboratory
environments. If the oxygen concentration in the environment was
constant while that of carbon dioxide varied, the curve was a rectangular hyperbola, being based on the law of mass action as applied to the
reversible reaction:

$$HBO_2 + CO \rightleftharpoons HbCO + O_2.$$

If, however, the dissociation curve was plotted for a situation where the
carbon monoxide concentration was constant while that of oxygen
varied, the curve deviated from a rectangular hyperbola. Evidently,
therefore, there were other processes occurring than just the reaction
mentioned above. This was not the first time that such deviations had
been observed and Hill (1910) had suggested aggregation among the
molecules of hemoglobin and oxyhemoglobin. However, the particular
mechanism of aggregation that Hill had suggested did not give the
correct rectangular hyperbola for the dissociation curve for hemoglobin
and carbon monoxide (at a constant oxygen concentration) though it

gave the correct curves when the carbon monoxide concentration was constant (while that of oxygen varied). Instead, Douglas *et al.* (1912) hypothesized the following sequence of reversible aggregation reactions for oxygen and hemoglobin:

$$HbO_2 + HbO_2 \rightleftharpoons Hb_2O_4$$
$$HbO_2 + Hb_2O_4 \rightleftharpoons Hb_3O_6$$
$$HbO_2 + Hb_2O_4 \rightleftharpoons Hb_4O_8$$
$$\cdot$$
$$\cdot$$
$$\cdot$$

and similar sequences for the interactions of hemoglobin with carbon monoxide and with itself. They then assumed that the law of mass action held for these reactions. From this assumption, and a few quite plausible ones about reaction rates, J. B. S. Haldane was able to calculate, using only some very basic algebra, the HbCO concentration at different oxygen concentrations. There was close agreement with the experimental data. What was crucial about this process was that the law of mass action was assumed to hold for molecules as large as hemoglobin and apparently did so.

It is clear that this type of physiological research satisfied Hopkins' dictum that biochemistry study dynamic processes, that is, the interaction of biological molecules. Moreover, the same can be said of Haldane's subsequent work in physiology at Oxford as well as at Cambridge. At the same time, such work definitely violated Hopkins' requirement that biochemistry be concerned with small molecules: the molecular weight of hemoglobin was known to be several tens of thousands. However, as has been observed above, Hopkins' purpose, in introducing this requirement, was to argue, mainly to the organic chemists whom he hoped to recruit, that biological molecules obeyed exact laws. If it turned out that large biological molecules obeyed exact laws anyway, and that these laws could be established by careful experimental techniques whose accuracy was beyond reasonable question, then the restriction to small molecules was unnecessary.[11] Consequently, Haldane's training in physiology met the three most important requirements that Hopkins felt important for biochemistry. The type of physiological work he had been doing was "dynamic." The molecules studied obeyed exact laws. Moreover, the gas apparatus devised by J. S.

Haldane for these experiments, and with which the younger Haldane was thoroughly familiar — in fact, according to legend, it is the only experimental apparatus with which he ever became remotely competent — were known to be remarkably accurate.

Finally, Haldane's interest in biology was above question — he was already emerging as an important geneticist. Thus, Hopkins' fourth requirement was met. When these factors are considered together with Haldane's reputation for brilliance and teaching skill, which were both in ample evidence at Oxford, and Hopkins' lack of skilled assistants who could teach (a point about which he complained quite bitterly in a letter to Fletcher (Hopkins 1919)), it does not even seem particularly perplexing, let alone paradoxical that Hopkins should have offered Haldane the new Readership at the Dunn Institute.

3

That Haldane's attention, in his "new" field of biochemistry, should focus on enzymes was perhaps inevitable. As Kohler (1973) has persuasively argued, the study of enzymes was central to the newly emerging independent discipline of biochemistry. Moreover, Haldane's teaching responsibilities, right from 1923, required delivering a course of lectures on enzymes.[12] Hopkins, too, must have emphasized the importance of enzymes. If Haldane's own words much later are reliable, or his publications accepted as sufficient evidence, he does not seem to have considered enzymes as a topic of research before. As he recalled, more than thirty years later: "Until I came to work in [Hopkins'] laboratory I had only used enzymes as analytical tools and had not thought about them for five consecutive minutes. Hopkins convinced me that they were a central topic in biochemistry (1965, vi)." Haldane's claim, here, can probably be taken at face value. Though Hopkins himself did not generally work with enzymes, references to them are scattered throughout his writings from this period. Further, several workers in his laboratory, including Quastel and Dixon, were carrying out important experimental work on enzymes.

Once Haldane's attention was focused on enzymes, it is also natural that his efforts would concentrate, or at least his mark would be felt, on enzyme kinetics. His lack of experimental knowledge of enzymology, and the difficulty of mastering the techniques involved, precluded experimental work unless he were to have taken it as his major commitment which, evidently, he was not prepared to do. Only theoretical possibilities remained and, in 1923, the kinetics of enzymes was about the only aspect of its study that it was possible to investigate, at least in part, theoretically.[13] The chemical nature of enzymes was still unknown.[14] It was not even generally accepted that enzymes were a single substance. Even in 1922, Wilstätter (1922) maintained that enzymes consisted of an active group and a colloidal carrier as did Fodor (1926) in 1926. Only in 1922 did Wilstätter and his collaborators begin the serious purification of enzymes. The first enzyme to be crystallized was jack bean urease by Sumner (1926) in 1926 and it remained controversial that the crystal obtained was that of the enzyme.[15]

The kinetics of enzymatic reactions depended, obviously, on the manner in which an enzyme acted upon its substrate. By 1923, there was almost universal consensus that enzymes acted by union with the substrate though even as late as 1924, a few biochemists (see, for example, Barendrecht (1924)) maintained that enzymes acted by emitting radiation. Theories of enzyme action and kinetics on the hypothesis of union had been begun to be formulated as early as 1902. In that year Brown (1902) observed that the rate of hydrolysis of cane sugar by saccharase was proportional to the enzyme concentration but was largely independent of the substrate (that is, sugar) concentration over a wide range. To explain this observation Brown hypothesized a mechanism by which the enzyme first forms a compound with the substrate which subsequently breaks down into the enzyme and various reaction products. As long as the substrate concentration is high enough to saturate the enzyme, increasing the substrate concentration would not increase the reaction rate. It was exactly here that enzymatic reactions differed from ordinary chemical ones where, by the law of mass action, the reaction rate was proportional to each of the reactants. Should Brown's hypothesized mechanism be correct, as long as the substrate concentration was not high enough to saturate the enzyme, the reaction rate would be proportional to its concentration but would be independent of the concentration of the enzyme. Once the enzyme was saturated, the reverse situation held as noted above. Since, in most enzymatic reactions, only a tiny amount of enzyme was present to act as a catalyst for the substrate, the second situation was expected. Around the same time Henri (1903) independently hypothesized the same mechanism for that enzyme and gave a mathematical treatment of the process. No further theoretical progress seems to have been made until 1913 when Michaelis and Menten (1913) generalized Henri's theory and gave it a more systematic and detailed treatment.

Michaelis and Menten (1913) analyzed the hydrolysis of sucrose to glucose and fructose, with yeast invertase acting as the catalytic enzyme.[16] If E, S, G and F represent molecules of enzyme, sucrose, glucose and fructose, respectively, Michaelis and Menten assumed that the following two reactions occur:

$$E + S \rightleftharpoons ES$$
$$ES + H_2O \rightarrow E + G + F.$$

They assumed that the first reaction was always at equilibrium. Let e be the *total* molar concentration of enzyme, x that of sucrose where it is assumed that x is much greater than e, and p that of ES molecules. Then the concentration of E is $e - p$. If K_1 is the dissociation (equilibrium) constant, then

$$K_1 = x(e - p)/p.$$

Therefore,

$$p = ex/(K_1 + x).$$

If the concentration of water molecules was nearly constant for the hydrolysis, and v is the rate of hydrolysis, and k a rate constant, using the law of mass action,

$$v = kp.$$

Using the formula for p given above,

$$v = kex/(K_1 + x).$$

When the concentration of the substrate is large enough for the enzyme to be saturated with substrate, a limiting rate is reached, as Brown (1902) had noticed. Let this be V. Then,

$$V = ke.$$

Therefore,

$$v = Vx/(K_1 + x).$$

Several algebraic variants of this were known, right from its initial introduction, as the Michaelis—Menten equation. If v is considered as a function of x, this equation represents a rectangular hyperbola. V is proportional to the enzyme concentration and can, therefore, be calculated by observations of maximum reaction rates at various enzyme concentrations. K_1 is characteristic of an enzymatic reaction and, once V is known, can be calculated from the form of the x-v hyperbola.

4

It was the assumption that the first reaction, that is, the one between the enzyme and the substrate, was always at equilibrium that Haldane found objectionalbe. However, he was well aware, in 1925, that the Michaelis—Menten equation seemed to hold for numerous enzymatic reactions (Briggs and Haldane 1925, 338). In 1925 what Briggs and Haldane (1925) explicitly set out to do is to provide a modified theory that served as the basis for the Michaelis—Menten equation.[17] In the process of developing this theory, which was entirely based on considerations regarding reaction rates obtained using the law of mass action, they observed that the assumption of equilibrium is equivalent to an assumption that the rate of the reaction:

$$ES + H_2O \rightarrow E + G + F$$

is negligible in comparison to that of

$$ES \rightarrow E + S.$$

They then went on to observe that "[i]t is clear, that data as to the course of a reaction can give no indication of the ratio (1925, 339)." By 1930, however, Haldane's objections to the assumption of equilibrium had become much stronger. Not content in nothing that information regarding it could not be obtained from the experimental results, he found the assumption improbable. The objections were based primarily on numerical considerations:

It has been assumed above that the combination of enzyme and substrate is always in equilibrium, i.e., the velocites of formation and dissociation of their compound are infinite in comparison with that of its decomposition to form the products of the reaction. This is rather an improbable assumption, for it seems likely that an invertase molecule can invert about 2000 sucrose molecules per second at 15 °C, so that the half-period of the reaction $ES \rightarrow E + G + F$ is less than 0.0005 second. It seems a little rash to postulate half periods of 10^{-5} second or less for other reactions. For only 10^7

cane sugar molecules in a 0.1 N solution collide with a given point with the enzyme surface per second, and we cannot suppose that the orientation would always permit of union with the enzyme (1930, 40).

In 1925, Briggs and Haldane (1925) made no assumptions about the relations between the three rates. Let k_1, k_2 and k_3 be the reaction rate constants for the following three reactions, respectively:

$$E + S \rightarrow ES$$
$$ES \rightarrow E + S$$
$$ES \rightarrow E + G + F.$$

Assuming the same symbol conventions as above, the rate of change of the ES concentration, (dp/dt), is given by:

$$(dp/dt) = k_1 x (e - p) - k_2 p - k_3 p.$$

Briggs and Haldane (1925, 338) argued that the value of (dp/dt) must be very small except during the first instant of the reaction. This provides a rationale for assuming it to be equal to 0 once the reactions have begun. In 1930 Haldane simply observed: "But so long as the velocity of the reaction is constant, p is constant, and even when it is altering, the rate of change of p must be infinitesimal compared to that of x (1930, 40)." Assuming $(dp/dt) = 0$ gives:

$$p = ex/(x + (k_2 + k_3)/k_1).$$

This equation, since called the Briggs–Haldane equation, assumed the form of the Michaelis–Menten equation if $(k_2 + k_3)/k_1$ is replaced by K_1. Strictly speaking, however, the Briggs–Haldane equation only reduces to the Michaelis–Menten equation if k_3 is equal to 0. Then k_2/k_1 is the equilibrium constant for the dissociation of ES, that is, it is K_1.[18]

In effect, what Briggs and Haldane (1925) had achieved was the replacement of Michaelis's and Menten's generally unrealistic assumption of an equilibrium between enzyme and substrate by what has since come to be known as an assumption of a "steady state," that is, one in which the rate of production of the enzyme-substrate intermediate complex is equal to the rate of its destruction. The Michaelis–Menten theory is obtained as a special degenerate case when the limit $k_3 \rightarrow 0$ is taken. Physically interpreted, this limit holds only when the formation of the reaction products from the enzyme-substrate complex ($ES \rightarrow E + G + F$) is much slower than its dissociation into enzyme and substrate ($ES \rightarrow E + S$) and it was exactly this situation whose plausiblity Haldane (1930) had called into question. The move from equilibrium kinetics to steady state kinetics has been the single most important conceptual development in the study of enzyme kinetics since Michaelis and Menten (1913). It remains Haldane's most important contribution to the study of enzymes.

Meanwhile, in 1930, Haldane conceded, however, that there were certain cases for which the Michaelis—Menten theory might suffice (1930, 41). These were cases such as that of yeast invertase for which, over a wide range of hydrogen ion concentration (pH) in the aqueous environment, though the rate constant for the ($ES \rightarrow E + G + F$) reaction, k_3, varied by a factor of 20, the measured value of K_1 appeared to remain constant. Unless, fortuitously, the other rate constants also varied in such a way to keep K_1 constant, the only reasonable conclusion appeared to be that, k_2 (which is the rate constant for the ($ES \rightarrow E + S$) reaction was much greater than k_3 as the Michaelis—Menten theory required. However, when the measured K_1 appeared to vary significantly with the hydrogen ion concentration (pH), he felt that some other theory such as the one proposed by Briggs and him was much more satisfactory.

5

Further mathematical development of the theory of enzyme kinetics for situations more complicated than the case considered above occupied Haldane in Chapter 5 of *Enzymes* (1930). Each of these situations was analysed by a modification of the simple mathematical treatment of that model. He considered seven types of cases:

(i) It was assumed that the products of the enzymatic reaction also interact with the enzyme (1930, 75—76). This interaction was inhibitive and the inhibition could be competitive or non-competitive. The mathematical treatment was restricted to the case of competitive interaction and was, essentially, identical to that already given by Michaelis and Menten (1913). Haldane found such a model plausible for saccharose hydrolysis.

(ii) It was assumed that the enzyme could be destroyed during the experiment (1930, 76—79). This could happen because of enzyme deactivation by heat in which case the destruction of the enzyme was independent of any interaction with the substrate. It could also happen that the substrate served to destroy the enzyme. Haldane provided a complete solution to these cases assuming Michaelis—Menten kinetics. The substrate could also potentially stabilize the enzyme. Haldane offered only a partial solution for the case.

(iii) It was assumed that the reaction betwen the enzyme-substrate complex and the reaction products was also reversible (1930, 80—83). In the first case this was a single-step process (like the one treated in detail above). If P represents the product of the reaction, the total reaction can be represented:

$$E + S \rightleftharpoons ES \rightleftharpoons E + P.$$

Haldane solved this case explicitly assuming steady-state kinetics. In the second case, it was assumed that the intermediate enzyme-substrate complex undergoes a transformation to an enzyme-product complex before the final dissociation into the enzyme and the product. This complication was introduced because the simpler case just considered did not seem to be in consonance with known relations for several enzymatic reactions. The second case can be represented:

$$E + S \rightleftharpoons ES \rightleftharpoons EP \rightleftharpoons E + P.$$

Once again, Haldane solved this case completely using steady-state kinetics. Moreover, these cases are particularly important because, implicit in Haldane's calculations (1930, 80—81), is a surprising relationship. If the maximum velocity and the equilibrium constant for the $(S \rightarrow P)$ reaction are V and K_1, respectively, and V' and K'_1 are the same quantities for the reverse reaction, then:

$$VK'_1/V'K_1 = C$$

where C is the product of the equilibrium constants of each of the intermediate stages. This fundamental relationship is generally known as the "Haldane relationship" though Haldane does not explicitly state it (see, e.g., Dixon and Webb 1979, 70).

(iv) It was assumed that the enzyme reacted with two substrates to form two different enzyme-substrate complexes (1930, 83—84). Each of these then reacted with the other substrate to form a more complicated intermediate complex at the next stage. Finally this complex gave rise to the reaction products.[19] Haldane solved this case assuming Michaelis—Menten kinetics.

(v) It was assumed that excessive substrate concentration inhibited enzyme activity (1930, 84—85). This was found to hold for some reactions catalyzed by lipases. Haldane solved this case using Michaelis—Menten kinetics.

(vi) It was assumed that substrates compete for the enzyme (1930, 85—88). Such behavior, too, was observed for lipases and Haldane solved this case once again assuming Michaelis—Menten kinetics.

(vii) It was assumed that there is a sequence of irreversible catalyzed reactions (1930, 88—91). Haldane did not attempt any general considerations of such cases but, assuming Michaelis—Menten kinetics, gave an analysis of some simple cases.

Besides this critical chapter, which contains the original mathematical results just noted, most of *Enzymes* consisted of a systematic survey and summary of the knowledge of enzymes as it stood in 1930, no doubt a reflection of the fact that the book arose out of a set of course lectures. However, in Chapter 9, he attempted some elementary applications of the kinetic theory to the interactions between enzymes and their substrates. This treatment is extended and somewhat refined in a later paper (Haldane 1931) using the more accurate data of Zeile and Hillström (1930) on the reaction of hydrogen peroxide and catalase. Qualitatively, the interesting results reached are that about 10^5 hydrogen peroxide molecules are catalysed by each catalase molecule per second and that an average atom, during its metabolic path in a cell, might interact with as many as 100 catalysts (1931, 567). After 1931 Haldane's interest in enzymes seems to have declined. In 1932 Haldane (1932a, b) presented cogent criticisms of the view championed by Haber and Willstätter (1931) that enzymes acted by initiating chain reactions, primarily arguing that whereas enzymatic reaction velocities were observed to be proportional to enzyme concentration, the chain theory implied that they would vary as the square root of that concentration.

In 1964, barely six months before his death, reflecting back on this early work, after noting how Hopkins convinced him that enzymes were central to biochemistry, he would claim: "My father, J. S. Haldane, had shown about 1910 that though haemoglobin is a large molecule, its reactions can be predicted from the laws known to hold for small

molecules. I had only to bring the ideas of these two great men together
to produce an account of enzyme action which, though sketchy, seems
to have been largely correct (1965, vi)." In this appraisal he might be
giving a little less credit than deserved to Brown (1902), Henri (1903)
and Michaelis and Menten (1913). After all, they, too had given at least
approximately correct accounts of enzyme kinetics based on the law of
mass action, then known to hold for certain only for small molecules.
However, these were all reactions of one large molecule, the enzyme,
with a small one and potentially, it can be argued, that it was not that
surprising that laws pertaining to small molecules would hold.[20] However, the same argument can also be brought to bear against Haldane
and Briggs (1925) and Haldane (1930). Thus, the claim that the
demonstration that the law of mass action also held for reactions
involving large molecules could not, ultimately, have played a critical
role in the elucidation of the laws of enzyme kinetics. However, in
claiming that he had J. S. Haldane's work in mind, Haldane (1965) was
probably not distorting how he thought of enzymatic reactions during
this period. *Enzymes* contains several comparisons to the work on
hemoglobin (see, e.g., 1930, 21) and explicit references to Douglas *et
al.* (1912; see, e.g., Haldane (1930, 44, 52)) and in the 1931 paper on
the molecular statistics of enzyme action (Haldane 1931), the various
parameters for enzymes are compared to similar ones for hemoglobin
(1931, 564—565).

6

Chapter 3 and 5 of *Enzymes* contain the entire theory of the kinetics of
enzymes that Haldane developed, that is, steady state kinetics. It is
instructive and important to note that the style of investigation is
quintessentially Haldane's in four critical ways. *First*, a simple model is
proposed and analyzed in detail. *Second*, the model and its analysis are
mathematical, though the mathematics remains very elementary. *Third*,
complicating factors are introduced as modifications of this simple
model which remains clearly in the background. *Fourth*, all complicating factors are introduced precisely in order to make the initial
model more "realistic" in the sense that they are motivated by a desire
to incorporate other mechanisms which have empirical support. There
is thus strong desire to relate theory to experiment.
    That this style is quintessentially Haldane's is best seen by noting

that it mimics the style of the seminal seriwa of papers on the mathematical theory of natural and artificial selection that Haldane was writing during this period. Chapter 5 of *Enzymes* is entitled 'The Course of Enzymatic Reactions and Its Mathematical Theory'; the first part of that series is called 'A Mathematical Theory of Natural and Artificial Selection' (Haldane 1924). This part simply presents the theory in its simplest form; the second part incorporates partial self-fertilization, inbreeding, assortative mating, and selective fertilization (1924a); mutation is incorporated much later (1927); then came isolation (1930a), and so on. The method is exactly the same as in enzyme kinetics. A simple model is introduced, analyzed mathematically using quite simple mathematics. Then complications are introduced as modifications of this simple model. Finally, the complications are introduced in order to make the model more "realistic"; those mechanisms are introduced that are known to occur in actual populations. In fact, as Allen (1978, 138) has noticed, a major difference between Fisher and Haldane, in their approach to evolutionary theory was that, whereas Fisher ignored specific factors operating in actual populations in order to simplify his models which he then developed with considerable mathematical elegance, Haldane concentrated on incorporating those factors so as to relate theory to experimental data. Exactly the same style was operative throughout Haldane's work on enzymes.

7

Meanwhile, in 1927 Haldane had accepted a part-time position as the director of genetical work at the John Innes Horticultural Institute at Merton. This institute had been established in 1909 from a bequest of £300 000 left by the late John Innes for the establishment of a technical school of horticulture.[21] Thanks to the intervention of several civil servants, especially from the Board of Agriculture, the use of the Innes bequest was reinterpreted to put an accent on research on plant-breeding. England's premier Mendelian geneticist, William Bateson, then facing an uncertain future at Cambridge, was offered the post of Director. Bateson accepted and research at John Innes, right from the beginning, was quite critically concentrated on the genetics of plants. However, Bateson died in 1926 and there was a strong possibility that Bateson's research program would soon dissipate (Darlington 1968; Harrison *ca.* 1980). The new Director, A. D. Hall, had no expertise in genetics.[22] Hall, however, was well aware of this limitation and, on Julian Huxley's advice, offered Haldane the position of Officer-in-Charge of genetical research. Haldane accepted apparently under a clear understanding that Hall would retire soon and that Haldane would succeed him (Harrison *ca.* 1980). According to Darlington (1968), who was already on the staff of John Innes at that time, Haldane's appointment prevented the collapse of the research program in genetics that Bateson had initiated.

Haldane served at John Innes for a decade (1927—1937), well past his tenure at Cambridge. Along with several collaborators who did the experimental work, he investigated the theoretical genetics of plants. In *Antirrhinum majus*, for example, he deduced from data previously collected at John Innes by I. B. Sutton and A. E. Gairdner that it included a case of linked balanced lethal factors, that is, factors which are lethal in the homozygote (Gairdner and Haldane 1929).[23] With D. De Winton he found a clear correlation between the presence of certain alleles in *Primula sinensis* and the presence of corresponding anthocyanin pigments suggesting that each allele was responsible for the presence of a particular pigment (De Winton and Haldane 1933). Later, they reported linkage in the same plant (De Winton and Haldane 1935). The former work was done in collaboration with R. Scott-Moncrieff whom Haldane had brought to John Innes from the Biochemical Laboratory at Cambridge. During the 1930's Scott-Moncrieff, W. J. C. Lawrence and, later, J. R. Price established the genetics of anthocyanin pigmentation in plants thereby paving the way for biochemical genetics, as most successfully established by G. W. Beadle and E. Tatum (1941). Work on anthocyanin at John Innes, however, had long preceded Haldane and Scott-Moncrieff.

8

By 1914, Muriel Wheldale, working under the direction of Bateson at John Innes had shown that the variation of anthocyanin pigments in plants might well be of great genetic interest.[24] Wheldale had worked with Bateson at Cambridge since 1902 and (along with Erwin Bauer in Germany) was uniquely ahead of her time by coupling breeding experiments on plants with chemical analysis of pigments. Bateson persuaded Wheldale to join the staff of John Innes in 1911 but she never became truly part of Bateson's group. Eventually she returned to Cambridge as a Fellow of Newnham College. However, as early as 1916, Wheldale had shown remarkable insight (perhaps exceeded only by Garrod (1909)) into the relation between genes and physiological factors and had written:

Herein lies the interest connected with the anthocyanin pigments. For we have now, on the one hand, satisfactory methods for the isolation, analyses and determination of the constitutional formulae of these pigments. On the other hand, we have the Mendelian methods for determining the laws of their inheritance. By a combination of these two methods we are within reasonable distance of being able to express some of the phenomena of inheritance in terms of chemical composition and structure (1916; as quoted in Scott-Moncrieff 1981).

At that time these hopes were overly optimistic. Not enough was known about the isolation of the anthocyanin pigments from plants to correlate them to Mendelian genes with success.

By the mid-1920's this situation had changed somewhat, thanks to the chemical methods developed by Wilstätter and others. Wheldale then advised Scott-Moncrieff, whose research she supervised at Cambridge, to tackle the problem anew (Scott-Moncrieff (1981, 126)). Haldane provided the necessary link to a genetical laboratory whereby the biochemistry of Cambridge could be coupled to Mendelian breeding experiments at John Innes. In 1929 Haldane established contact between Scott-Moncrieff and Lawrence (at John Innes). Together they began a biochemical survey of related genotypes of *Antirrhinum majus, Primula polyanthus* and other plants containing anthocyanin pigments.

That Haldane, with interests in both genetics and biochemistry, should have encouraged such research is hardly surprising. Moreover, as early as 1920, Haldane had speculated: "The precise nature of [the genes'] activity is uncertain, but in some cases we have very strong evidence that they produce definite quantities of enzymes, and that the members of a series of multiple allelomorphs produce the same enzymes in different quantities (1920, 10)." He had also realized the significance of the work of Garrod (1909) in providing a link between genetics and metabolism. Wheldale's suggestion to Scott-Monscrieff, therefore, generated his immediate support and enthusiasm. His own position, as a member of both the Biochemistry Laboratory at Cambridge and John Innes provided a unique opportunity for the initiation of an inter-disciplinary research program that would investigate as fundamental a problem as that of the mode of action of the genes through definite chemical products that mediated the metabolic life of organisms. To Scott-Moncrieff, however, belongs the most credit for establishing this research program with success. Though Haldane supervised her research he did not perform any of the experiments. He did, however, help in their analysis at the final stages.

9

By the time Scott-Moncrieff began working in earnest at John Innes, she had already received her "PhD (Titular) Certificate" from Cambridge in 1930. Cambridge, in those days, did not deign to give women actual degrees — a "Title of a Degree" was the best that a woman could expect.[25] She had also achieved a major diplomatic success. She had succeeded in ensuring a collaboration between the organic chemistry laboratory of R. Robinson at Oxford and the geneticists at John Innes. While Scott-Moncrieff had isolated several natural anthocyanins from plants Cambridge, Robinson and his laboratory at Oxford were synthesizing them (Scott-Moncrieff 1981). Even more important, Robinson's laboratory had devised qualitative tests that could detect minute quantities of the pigments (Robinson and Robinson 1931). This ability was essential for the detection of possible genetic dfferences between plants whose flowers did not vary sufficiently in color to make visual recognition reliable. However, Robinson's opinion of genetic methods was not high: "Like cracking a walnut with a sledge hammer," he had confided to Scott-Moncrieff (1981, 142).

Robinson had, however, corresponded with Haldane (see, e.g., Haldane (*ca.* 1928)). At stake was the cause of pigment differences in plants: Robinson speculated that colloidal differences were involved, Haldane suggested pH differences instead.[26] In any case, Robinson invited Scott-Moncrieff to collaborate with his laboratory while working under Haldane. It require considerable tact for Scott-Moncrieff to hold the collaboration together. But, out of it, Robinson lost his distrust of the geneticists. Out of it also emerged the new field of chemical genetics. In the 1930's two major discoveries were made at John Innes. *First*, it was discovered that changes in flower color variation were either controlled by single-locus differences or by a set of specific genes. The biochemical interrelationship was as exact as the genetic. *Second*, biochemical effects were found to be uniform in a species.

Robinson aptly summarized the new situation in 1936, in several lectures on the formation of anthocyanins in plants (whose contents were subsequently published as a single piece in *Nature*):

Lawrence and Scott-Moncrieff in their studies of the garden *Dahlia* have founded what is, in effect, a new subject of chemical genetics. It is not suggested that genetics and chemistry have never previously been considered together, but this pioneering work shows not only how to apply chemical methods of examination to the routine of genetic work but also, which is much more novel, how to interpret the results so as to throw light on the actual mechanism of synthesis in the plant (1936, 172).

In 1937, however, Scott-Moncrieff left John Innes to marry and live in India. The research program there was continued by Lawrence and Price, the latter having come directly from Robinson's laboratory at Oxford. Haldane, too, left John Innes in 1937 after a bitter dispute with Hall.[27]

10

There are scattered references to biochemistry in Haldane's *Causes of Evolution* (1932) and an implicit assumption throughout that any account of the action of genes must be consistent with what might transpire at the underlying biochemical level. Biochemical differences could well be one of the ways in which species could be defined, Haldane argued (1932, 63) and even went on to claim that it was known for certain that some genes acted by inhibiting a specific enzyme (1932, 70). The non-additive nature of the simultaneous action of several genes was obvious if genes acted chemically (1932, 96). That apparently continuous variation between individuals in a species was deceptive, if the atomism inherent in Mendelism is true, was also necessary on "any chemical theory of the nature of genes (1932, 104)." And, finally, he stated explicitly: "To my mind it is probable that every gene produces a definite chemical effect, but we are far from being able to prove this as yet (1932, 115). In striking contrast, in Fisher's *Genetical Theory of Natural Selection* (1930), there is no mention of biochemistry at all.

The most succinct and penetrating discussion of Haldane's views on the relation between genes and their specific chemical products, especially enzymes, is contained, however, in a long unpublished manuscript from about 1931 (Haldane *ca.* 1931). Haldane began by noting that, thanks to the work of the Morgan school, it was known that genes were localized on chromosomes, that they were doubled each generation, that recessive genes were often variants of dominant ones and that the molecular weight of a gene could be estimated to be under 10^5 (daltons). He proceeded to construct a biochemically viable account of genes that explained these facts. "To a biochemist," he claimed, "it is obvious that morphlogical differences [between different allelomorphs] are in all probability due to chemical differences (ca. 1931, 338)." Therefore, to understand morphological differences, the chemical differences needed to be studied. Unfortunately, such studies were scanty. It was known — from the sort of work summarized in the last two sections — that in some cases certain genes suppressed the formation of a definite pigment. In some of these cases he speculated that the genes might be acting (directly or indirectly) through the formation of an acid, emphasizing once again his view that pH differences were crucial to coloration in flowers. He then listed other possible mechanisms and went on to make a remarkable set of observations: "How do these genes carry out the chemical reactions in question, probably reduction on the one hand, methylation on the other? In neither case can we answer, though in each case enzymatic catalysis is quite plausible . . . [W]e can point to quite definite enzymes in a variety of cases (ca. 1931, 341)." As evidence he presented not only the cases of pigmentation in plants but also the work on melanin formation on rodents that had also been carried out at Cambridge.[28] "The interaction," he emphasized, "is not between the genes themselves but between their products, and is a matter of biochemistry. . . . The enzyme is a product of the gene, not the gene itself (ca. 1931, 342—344)." Besides enzymes, Haldane further argued, "genes can produce at least one other class of specifically active colloids, namely immune bodies (ca. 1931, 344)." Examples included the agglutinogens of red blood corpuscles. Finally, he observed: "All these and many other results can be explained on the following hypothesis. 'A gene of any type produces a specific and generally colloidal substance. The effects of genes in combination are determined by the interaction of these substances with each other and the remaining cell constituents (ca. 1931, 345—346).'"[29]

    In a concluding and self-consciously speculative section of the manuscript Haldane turned from the nature of gene action to the nature of the gene itself. Observing that since irreversible changes (such as mutations) in the genetic material (the chromatin) were scrupulously copied, he argued that the chromatin must be of a simpler character than the cytoplasmic materials to make such copying possible. He went on to observe: "If we are to picture the copying process it must be conceived somewhat as follows. A certain molecular grouping only the molecule thick must attract dissolved molecules from solution so that a similar layer of molecules is formed on it. Such a process is remarkably like crystallization, like attracting like (*ca.* 1931, 359)." However, unlike crystallization, the process stopped after the next layer was formed. The ability of a gene to direct the synthesis of specific chemical products resided in the morphological arrangement of the constituent parts or units of the chromatin. These units, according to Haldane, could be whole protamine or nucleic acid molecules but were likely to be smaller entities such as nucleotides and peptides. Particularly presciently, he observed that a very large number of arrangements of such units were easily conceivable and construed most mutations as changes in such arrangements. These speculations, and the manuscript itself, were never completed and published. Had this manuscript been published it would have done much to justify Haldane's later, and persistent, insistence that he was one of the major precursors of the "one gene — one enzyme" hypothesis that played so major a role in the early history of molecular biology.[30]

11

By 1933, however, Haldane had abandoned biochemistry proper for all practical purposes though the interest in biochemical genetics would remain throughout his life. Instead, he had finally decided to make genetics the center of his interests. This decision to concentrate solely on genetics was accompanied by a corresponding switch in academic posts. In 1933 he became Professor of Genetics at University College, London, a post that was created there following Pearson's retirement from the Chair of Eugenics and the subsequent reorganization of the Department of Applied Statistics (Clark 1969, 92—93). Though an official statement from University College claimed that Haldane's "great knowledge of Biochemistry will enable him to introduce into Genetics, which has heretofore been purely morphological in nature, those physiological conceptions which alone can lead to an understanding of the mode of action of the genes on which the hereditary transmission of structure rests (Clark 1969, 93)," Haldane's work at University College would scarcely touch biochemistry. No collaboration of the sort between Cambridge and John Innes would emerge though Scott-Moncrieff, Lawrence and others would pursue biochemical genetics at John Innes under Haldane's general direction unitl 1937.

The decision to concentrate on genetics was undoubtedly motivated by the acclaim that the mathematical theory of natural and artificial selection had received: in 1932 Haldane was elected to Fellow of the Royal Society. It had other rewards, too. From Cambridge, Punnett wrote to Haldane: "Now that you have definitely cast in your lot with Genetics (and I am very glad that you have so decided) I want to ask you whether you would care to help me with the Journal of Genetics (Punnett 1933)." Punnett offered an equal editorial role and responsibility and a meager financial remuneration. "I should like the John Innes to be represented on the Journal," Punnett also observed. "For they provide more new material than any other place. . . . I have my eye on the time when I retire (not so very far away) and have to come to a decision as to the way in which the Journal is to be carried on." Haldance accepted. His relations with John Innes ended in 1937. However, the association with the *Journal of Genetics* would remain to the end of his life (though interrupted by World War II) when he would take the *Journal* with him to India.

Boston University

NOTES

[1] Thanks are due to James F. Crow, Rafael Falk and Alfred I. Tauber for comments on an earlier draft of this paper. This paper would not have been possible without the assistance of Ms. G. Furlong and the staff of the archives of University College Library, London; Mrs. R. D. Harvey, Archivist, John Innes Institute, Norwich; the archival staff of the National Library of Scotland; the archival staff of Cambridge University Library, and the encouragement and help of James F. Crow, John Maynard Smith, Graeme Mitchison and Naomi Mitchison. The work reported here was supported by an archival research grant from the American Philosophical Society. This is Contribution No. BTBG 90-6 from the Theoretical Biological Group, Boston Center for the Philosophy and History of Science.
[2] The only exceptions seem to be Caspari (1968), Wurmdser (1968) and Scott-Moncrieff (1981).
[3] Indeed, Haldane's materialism always remained distinct from the mechanistic conception of materialism championed by Hopkins (see, for example, Haldane (*ca.* 1938)). In the late 1930's he endorsed dialectical materialism shortly before he joined the

Communist Party. He seems to have abandoned any strict commitment to dialectical materialism only in the 1950's — at least he stopped proselytizing or writing about it — as he moved away from the Communist Party but never explicitly abandoned materialism altogether. However, at no stage, did this materialism become the simple "mechanism" of committed reductionists.

[4] Bibliographical details about Hopkins are from Baldwin (1972).

[5] The strain of such disparate responsibilities was such that it led to a physical breakdown in 1910.

[6] For details, see Kohler (1973).

[7] Details of the establishment of biochemistry at Cambridge and, in particular, of the Dunn Institute of Biochemistry are found in Kohler (1978).

[8] This paper will be discussed in some detail below.

[9] Kohler (1973) has argued that the new study of enzymes was essential to the emergence of biochemistry as will be noted below in the text. However, though this observation does show something about the conceptual structure of the new discipline, it does not constitute definitive evidence against the possibility that a new concentration on enzymes could have been accommodated as a sub-discipline within the physiological profession.

[10] See, especially, his letter to Fletcher from June 9, 1919 where he asks for the latter's help in securing funding for a department and outlines his difficulties and desires (Hopkins 1919).

[11] Further evidence that large biological molecules obeyed simple laws was also available from the work of Brown (1902), Henri (1903) and Michaelis and Menten (1913) which will be discussed in the text. However, these concerned the interaction of a small molecule (the substrate) with a large one (the enzyme). The aggregation process considered by Douglas et al. (1912), however, truly concerned interactions of large molecules with each other.

[12] See the Preface of Haldane (1930).

[13] An unfortunately short, but accurate, account of the history of the study of enzymes is Dixon (1970). More detail is found in Fruton (1972).

[14] Even in 1930 Haldane would observe in Enzymes: "The attempts to purify enzymes have led to no definite conclusion as to their chemical nature. Except Sumner's urease, none of the most highly purified preparations appeared to be proteins (1930, 179)."

[15] See, for example, Haldane's qualified observation: "Sumner believes that the crystalline protein is the enzyme (1930, 170). See, also, Dixon (1970, 24).

[16] For the sake of consistency, the notation followed throughout this section is basically that of Haldane (1930) and not that of Michaelis and Menten (1913) or of Briggs and Haldane (1925).

[17] Yet another alternative derivation of the Michaelis—Menten equation, but on the assumption that the formation of ES is not reversible was due to Van Slyke and Cullen (1914). It does not seem to have had much impact.

[18] It is $(k_2 + k_3)/k_1$ that is customarily called the Michaelis constant now.

[19] These are now called "random order equilibrium pathways" (Dixon and Webb 1979, 88).

[20] This is especially likely if it is assumed, as it often was, that the enzyme-substrate interaction was one of adsorption of the substrate on the enzyme surface rather than something like chemical union.

[21] For details of the history of the foundation of the John Innes Horticultural Institute and its early direction under Bateson, see Olby (1989).

[22] According to Lawrence (1980, 30), Haldane "had greatly hoped to succeed Bateson but his political affiliations were distasteful to the governing body of the Institute." This might well be true, as Haldane's politics had slowly but surely moved to the left during the Cambridge years. However, there is no independent evidence in support of this assessment.

[23] Most of these collaborators were women who had been employed by Bateson since their labor was cheaper than that of men (Harrison ca. 1980).

[24] Biographical details about Wheldale are from Olby (1989) and Scott-Moncrieff (1981).

[25] For a discussion of the reluctance of Cambridge to recognize women as academic equals of men, see Howarth (1978).

[26] It turns out both were correct. See Haldane (1954, 55) for details.

[27] Though the early history of "chemical genetics" briefly recorded here seems not to be well-known, Beadle clearly recognized and acknowledged its importance in motivating the work of Beadle and Tatum (Beadle 1974).

[28] The work was mainly that of Huia Onslow. See Onslow (1924) for detail.

[29] Haldane went on to use this specific biochemical conception of gene action to argue against Fisher's theory of dominance. While that argument is long and complicated, relying on the law of mass action holding between a gene and its chemical product, Haldane's attempt to find such an explanation underscores the importance he placed on giving biochemical accounts of genetic phenomena.

[30] See Scott-Moncrieff (1981) and Haldane (1954) for evidence of Haldane's insistence about his role in these developments.

REFERENCES

Allen, G. (1978), *Life Science in the Twentieth Century*, Cambridge: Cambridge University Press.

Baldwin, E. (1972), 'Frederick Gowland Hopkins', *Dictionary of Scientific Biography* 6: 498—502.

Barendrecht, H. P. (1924), 'Saccharase und die Zweite Wirkungsart der Wasserstoffionen', *Biochemische Zeitschrift* 151: 363—370.

Beadle, G. W. (1974) 'Recollections', *Annual Review of Biochemistry* 43: 1—13.

Beadle, G. W. and Tatum, E. (1941), 'Genetics and Metabolism in Neurospora', *Proceedings of the National Academy of Sciences (USA)* 27: 499—506.

Briggs, G. E. and Haldane, J. B. S. (1925), 'A Note on the Kinetics of Enzyme Action', *Biochemical Journal* 19: 338—339.

Brown, A. J. (1902), 'Enzyme Action', *Transactions of the Chemical Society* 81: 373—388.

Caspari, E. (1968), 'Haldane's Place in the Growth of Biochemical Genetics', in Dronamraju (1968, 43—50).

Clark, R. W. (1969), *J B S: The Life and Work of J. B. S. Haldane*, New York: Coward-McCann.

Clark, R. W. (1972), 'John Burdon Sanderson Haldane', *Dictionary of Scientific Biography* 6: 21—23.

Darbishire, A. D. (1904), 'On the Result of Crossing Japanese Waltzing with Albino Mice', *Biometrika* 3: 1—51.

Darlington, C. D. (1968), 'Determined but Lonely', *Nature* 220: 933—934.

De Winton, D. and Haldane, J. B. S. (1933), 'The Genetics of *Primula Synensis* II: Segregation and Interaction of Factors in the Diploid', *Journal of Genetics* 27: 1—44.

De Winton, D. and Haldane, J. B. S. (1935), 'The Genetics of *Primula Synensis* III: Linkage in the Diploid', *Journal of Genetics* 31: 67—100.

Dixon, M. (1970) 'The History of Enzymes and of Biological Oxidation', in Needham (1970, 15—37).

Dixon, M. and Webb, E. C. (1979), *Enzymes*, 3rd ed. New York: Academic Press.

Douglas, C. G., Haldane, J. S., and Haldane, J. B. S. (1912), 'The Laws of Combination of Haemoglobin with Carbon Monoxide and Oxygen', *Journal of Physiology* 44: 275—304.

Dronamraju. K. R. (ed.) (1968), *Haldane and Modern Biology*, Baltimore: John Hopkins Press, 313—317.

Fisher, R. A. (1930), *The Genetical Theory of Natural Selection*, Oxford: Clarendon Press.

Fodor, A. (1926), 'Fermentwirkung und Wasserstoffionenkonzentration', *Kolloid-Zeitschrift* 40: 234—240.

Fruton, J. S. (1972), *Molecules and Life: Historical Essays on the Interplay of Chemistry and Biology*, New York: Wiley-Interscience.

Gairdner, A. E. and Haldane, J. B. S. (1929), 'A Case of Balanced Lethal Factors in *Antirrhinum majus*', *Journal of Genetics* 21: 315—325.

Garrod, A. E. (1909), *Inborn Errors of Metabolism*, London: Frowde, Hodder, and Stroughton.

Haber, F. and Willstätter, R. (1931), 'Unpaarigkeit und Radikalketten im Reaktions-mechanismus Organischer und Enzymatischer Vorgänge', *Berichte der Deutchen Chemischen Gesselschaft* 64: 2844—2856.

Haldane, J. B. S. (1920), 'Some Recent Work on Heredity', *Transactions of the Oxford University Junior Scientific Club*, Series 3, No. 1, pp. 3—11.

Haldane, J. B. S. (1922) [Letter to F. G. Hopkins, May 23, 1922], Cambridge University Library Archives, Cambridge, BCHEM 3/6.

Haldane, J. B. S. (1924), 'A Mathematical Theory of Natural and Artificial Selection. Part I', *Transactions of the Cambridge Philosophical Society* 23: 19—41.

Haldane, J. B. S. (1924a), 'A Mathematical Theory of Natural and Artificial Selection. Part II. The Influence of Partial Self-Fertilization, Inbreeding, Assortative Mating, and Selective Fertilization on the Composition of Mendelian Populations, and on Natural Selection', *Proceedings of the Cambridge Philosophical Society* 1: 158—163.

Haldane, J. B. S. (1927), 'A Mathematical Theory of Natural and Artificial Selection. Part V. Selection and Mutation', *Proceedings of the Cambridge Philosophical Society* 23: 838—844.

Haldane, J. B. S. (*ca.* 1928) [Draft of letter to R. Robinson, undated], Haldane Papers, National Library of Scotland, Edinburgh, No. 29546, ff. 191—193.

Haldane, J. B. S. (1929), 'The Origin of Life', *Rationalist Annual*, pp. 3—10.

Haldane, J. B. S. (1930), *Enzymes*, London: Longmans, Green and Company.
Haldane, J. B. S. (1930a), 'A Mathematical Theory of Natural and Artificial Selection. Part VI. Isolation', *Proceedings of the Cambridge Philosophical Society* **26**: 220—230.
Haldane, J. B. S. (1931), 'The Molecular Statistics of an Enzyme Action', *Proceedings of the Royal Society (London) B*: 559—567.
Haldane, J. B. S. (*ca.* 1931), 'A Biochemical View of Some Genetical Problems', Unpublished. Haldane Papers, Nation Library of Scotland, Edinburgh. No. 20580, ff. 36—65.
Haldane, J. B. S. (1932), *The Causes of Evolution*, London: Longmans, Green and Company.
Haldane, J. B. S. (1932a) [Discussion of Recent Advances in the Study of Enzymes and their Action], *Proceedings of the Royal Society B* **111**: 292—294.
Haldane, J. B. S. (1932b), 'Chain Reactions in Enzymatic Catalysis', *Nature* **130**: 61.
Haldane, J. B. S. (1934), 'A Mathematical Theory of Natural and Artificial Selection. Part X. Some Theorems on Artificial Selection', *Genetics* **19**: 412—429.
Haldane, J. B. S. (*ca.* 1942), 'Why I am Cooperator [sic]', Unpublished Manuscript, Haldane Collections, University College Archives, London.
Haldane, J. B. S. (1954), *The Biochemistry of Genetics*, London: George Allen and Unwin.
Haldane, J. B. S. (1965), 'Preface to the Paperback Edition', in Haldane, J. B. S., *Enzymes*, Cambridge, MA: MIT Press, v—vi.
Haldane, J. B. S., Sprunt, A. D., and Haldane, N. M. (1915), 'Reduplication in Mice', *Journal of Genetics* **5**: 133—135.
Harrison, B. J. (*ca.* 1980). [Taped Interview with C. D. Darlington], John Innes Institute Archives, Norwich.
Henri, V. (1903), *Lois Générales de l'Action des Diastases*, Paris: Hermann.
Hill. A. V. (1910), 'The Possible Effects of the Aggregation of the Molecules of Haemoglobin on Its Dissociation Curves', *Journal of Physiology* **40**: iv—vii.
Hopkins, F. G. (1913), 'The Dynamic Side of Biochemistry', in Needham and Baldwin (1949, 136—159).
Hopkins, F. G. (1919) [Letter to Walter Fletcher, June 9, 1919], Reprinted (in part) in Kohler (1973, 352—354).
Hoppe-Seyler, F. (1877), 'Vorwort', *Zeitsschrift für Physiologische Chemie* **1**: 1.
Howarth, T. E. B. (1978), *Cambridge between Two Wars*, London: Collins.
Kohler, R. E. (1973), 'The Enzyme Theory and Biochemistry', *Isis* **64**: 181—196.
Kohler, R. E. (1978), 'Walter Fletcher, F. G. Hopkins and the Dunn Institute of Biochemistry: A Case Study in the Patronage of Science', *Isis* **69**: 331—355.
Lawrence, W. G. C. (1980), *Catch the Tide: Adventures in Horticultural Research*, London: Grower Books.
Michaelis, L. and Menten, M. L. (1913), 'Zur Kinetik der Invertinwirkung', *Biochemische Zeitschrift* **49**: 333—369.
Needham, J. (ed.) (1970), *The Chemistry of Life: Eight Lectures on the History of Biochemistry*, Cambridge: Cambridge University Press.
Needham, J. and Baldwin, E. (eds.) (1949), *Hopkins & Biochemistry: 1861—1947*, Cambridge: W. Heffer and Sons.

Olby, R. C. (1974), *The Path to the Double Helix*, Seattle: University of Washington Press.

Olby, R. C. (1989), 'Scientists and Bureaucrats in the Establishment of the John Innes Horticultural Institution under William Bateson', *Annals of Science* **46**: 497—510.

Onslow, M. W. (1924), *Huia Onslow: A Memoir*, London: E. Arnold.

Provine, W. (1971), *The Origins of Theoretical Population Genetics*, Chicago: University of Chicago Press.

Punnett, R. C. (1912) [Letter to J. B. S. Haldane, undated], Haldane Papers, National Library of Scotland, Edinburgh, No. 20546, f. 189.

Punnett, R. C. (1933) [Letter to J. B. S. Haldane, 6. 30. 1933], Haldane Papers, National Library of Scotland, Edinburgh, No. 20534, ff. 23—25.

Robinson, R. (1936), 'Formation of Anthocyanins in Plants', *Nature* **137**: 172—173.

Robinson, G. M. and Robinson, R. (1931), 'A Survey of Anthocyanins. I', *Biochemical Journal* **25**: 1687—1705.

Scott-Moncrieff, R. (1981), 'The Classical Period in Chemical Genetics. Recollections of Muriel Wheldale Onslow, Robert and Gertrude Robinson and J. B. S. Haldane', *Notes and Records of the Royal Society of London* **36**: 125—154.

Sumner, J. B. (1926), 'The Isolation and Crystallization of the Enzyme Urease', *Journal of Biological Chemistry* **69**: 435—441.

Van Slyke, D. D. and Cullen, G. E. (1914), 'The Mode of Action of Urease and of Enzymes in General', *Journal of Biological Chemistry* **19**: 141—180.

Werskey, G. (1972), *The Visible College: The Collective Biography of British Scientific Socialists of the 1930s*, New York: Holt, Rinehart and Winston.

Wheldale, M. (1916), *The Anthocyanin Pigments of Plants*, Cambridge: Cambridge University Press.

Wilstätter, R. (1922), 'Über Isolierung von Enzyme', *Berichte der Deutschen Chemischen Gesellschaft* **55**: 3601—3623.

Wright, S. (1931), 'Evolution in Mendelian Populations', *Genetics* **16**: 97—159.

Wright, S. (1968), 'Contributions to Genetics', in Dronamraju (1968, 1—12).

Wurmser, R. (1968), 'Haldane as I Knew Him', in Dronamraju (1968, 313—317).

Zeile, K. and Hillström, H. (1930), 'Über die Aktive Gruppe der Leberkatalase', *Zeitschrift für Physiologische Chemie* **192**: 171—192.

JAMES F. CROW

H. J. MULLER'S ROLE IN EVOLUTIONARY BIOLOGY

To many an unsophisticated human being, the universe of stars seems only a fancy backdrop, provided for embellishing his own and his fellow creatures' performances. On the other hand, from the converse position, that of the universe of stars, not only all human beings but the totality of life is merely a fancy kind of rust, afflicting the surfaces of certain lukewarm minor planets. However, even when we admit our own littleness and the egotistical complexion of our interest in this rust, we remain confronted with the question: What is it that causes the rust to be so very fancy?

H. J. Muller (1955)

It is customary, in discussions of modern theories of evolution, to pay tribute to the three giants — R. A. Fisher, J. B. S. Haldane, and Sewall Wright. It is more than customary, it is *de rigueur*. Some of their contemporaries who also had a large influence through research or writing were G. G. Simpson, Th. Dobzhansky, S. S. Tschetverikoff, E. Mayr, J. S. Huxley, N. W. Timofeeff-Ressovsky, A. H. Sturtevant, and G. L. Stebbins. And of course a subject so central to biology has had many other contributors. Yet, I have omitted one important name.

H. J. Muller is widely regarded as the greatest geneticist in the first half-century of the science.[1] To cite but one example, his "Pilgrim Trust Lecture" (Muller, 1947) gave a remarkably perceptive explanation for bacterial DNA transformation and foretold "the coming chemical attack on the gene". His contributions to evolution are less well recognized. Yet they are substantial, and I think Muller should be regarded as one of the leading evolutionary thinkers. It is some of his contributions to evolution that I should like to discuss.[2]

Muller's view of evolution was remarkably broad. Only Haldane, who wrote technical articles on the origin of the universe and life (e.g., Haldane, 1945), rivals him in this regard. Muller regularly taught a course in evolution, which started with the origins of the universe and of the solar system and ended with science fiction-like speculation about, and hopes for, the future of human evolution.[3]

Muller and Sturtevant were fellow students in T. H. Morgan's laboratory at Columbia University. Both already had an interest in evolution. Sturtevant was the first to study different species of Drosophila by both

83

Sahotra Sarkar (ed.), The Founders of Evolutionary Genetics, 83—105.

taxonomic and breeding methods. He discovered, among other things, that *Drosophila melanogaster* and *D. simulans* would hybridize, but that the hybrids were sterile.[4] Sturtevant initiated the study of natural populations of other Drosophila species, a subject taken up by Th. Dobzhansky (1937) and pursued with his characteristic vigor. This work constitutes a major leap in the experimental study of evolution. It was Muller, however, who followed up Sturtevant's study and devised a clever way to bypass the sterility of the hybrids (Muller and Ponte-corvo, 1940). I shall return to this later.

In one respect, Muller and Sturtevant stand in sharp contrast. Sturtevant's published experiments were remarkable for their clarity of purpose, clean design, and unambiguous conclusions. A Sturtevant experiment didn't have to be repeated. The results were reported in clear, terse prose, and that was all. There was a minimum of discussion, and what discussion there was adhered strictly to the point. Muller, in contrast, wrote expansively, described everything meticulously, and — especially — discussed all the implications he could think of (and he could think of a lot). Not only did he have a multitude of ideas, he made a point in his writing of including as many of them as he could legitimately drag in.[5]

Characteristically, Sturtevant insisted that his correspondence be destroyed. He said that what was important and clearly thought out was in his published papers, and he saw no reason why anyone should care about his half-baked, preliminary ideas. Muller, in contrast, saved *everything*. Even travel schedules and receipts are dutifully maintained in the Lilly Library at Indiana University, which houses Muller's files.[6]

I mention Muller's habit of writing down every idea, and if possible including it in a paper, for a specific reason. It is easy to document Muller's ideas, for so many of them are in writing. It is quite possible that many of these also occurred to others — Sturtevant or L. J. Stadler, for example — but were not recorded. Thus, if we had a full record of everyone's thoughts, Muller's reputation as the idea-man of genetics might be diminished. Nevertheless, I suspect that even if this were possible, Muller would still be the leader. The ideas that he wrote down are far more impressive than would be a complete transcription of the thoughts of most scientists.

THE GENE AS THE BASIS OF LIFE

One idea, first clearly articulated by Muller, is now a commonplace and a part of conventional wisdom. Its origins are forgotten. We fail to realize how revolutionary it was to biologists three quarters of a century ago. I refer to Muller's view of the gene as the basis of life (Muller, 1922, 1929, 1932).

By 1920 the studies of the Drosophila group at Columbia University had established most of the phenomenology of mutation: that the gene was equally stable before and after mutation; that gene mutation occurred independently of other genes, and was not contaminated by them, even by the allelic gene; that the mutation process was independent of external environment;[7] and, most important, that mutations were random with respect to their phenotypic effects and therefore only a small fraction were beneficial.

Although Muller, along with others, emphasized the importance of the accurate self-replication of the gene, he was the first to emphasize that it is the ability to replicate mistakes that is the important property making evolution, and indeed life itself, possible. Once an entity exists that can faithfully replicate its (relatively rare) errors, natural selection can go to work to produce higher levels of adaptation and greater complexity (Muller, 1929).

As I said, this view of gene-primacy is now standard. I think Muller might also be regarded as the first to introduce the selfish gene concept, although the words are not his. That the first meaningful form of life was a primitive gene was very much part of Muller's thinking. Recent work on the catalytic function of RNA (see Gilbert, 1986) adds to the relevance of experimental studies of nucleic acids in the origin of life.

But the point that Muller most often emphasized is that it is the properties of making errors (which is, of course, not remarkable) and faithfully copying these errors (which *is* remarkable) that provide the basic requisite of life, that is to say, a system capable of undergoing evolution by natural selection. Needless to say, Muller greeted the Watson—Crick model of DNA structure with immense enthusiasm. DNA was a simple (though enormous) molecule that did just what he thought the gene ought to do.

I should like to quote what is probably Muller's most famous statement, a prophetic one made in 1922 when bacterial viruses were very poorly understood.

If these . . . bodies [bacteriophage] were really genes, fundamentally like our chromosome genes, they would give us an utterly new angle from which to attack the gene
problem. They are filterable, to some extent isolable, can be handled in test tubes, and
their properties, as shown by their effects on the bacteria, can then be studied after
treatment. It would be very rash to call these bodies genes, and yet at present we must
confess that there is no distinction known between the genes and them. Hence we
cannot categorically deny that perhaps we may be able to grind genes in a mortar and
cook them in a beaker after all. Must we geneticists become bacteriologists, physiological chemists and physicists, simultaneously with being zoologists and botanists? Let
us hope so.[8]

An idea that is usually credited jointly to Muller and C. B. Bridges is
that new genes don't arise *de novo* but come from pre-existing genes. A
classical dilemma of Mendelian genetics is that when a gene mutates,
and possibly gains a new, useful function, it loses the old one.
Mendelism and mutation can substitute, Muller said, but they can't add.
Muller pointed out that one way the organism can have its cake and eat
it is somehow to duplicate the gene. Then one copy can mutate to
produce a new function while the other retains the old one. Thus the
organism acquires new genes, gains new functions, and increases in
complexity.

Although the idea must have been around for some time, it was not
until 1936 that Bridges and Muller — independently, Bridges in
California and Muller in Russia — showed that the "Bar eye" phenotype in Drosophila was the result of a gene duplication, and arose by a
mistake in crossing over. The idea was now proven, at least for this
case. Subsequent history has provided abundant examples and many
additional mechanisms, especially transposons and retroviruses.

So, we have the answer to the question of where new genes come
from: from old ones. In Muller's (1936) words, there is "no reason to
doubt the application of the dictum 'all life from preexisting life' and
'every cell from a pre-existing cell' to the gene: 'every gene from a
pre-existing gene'. We need at present make an exception here only of
those very special conditions under which life itself, as a naked gene,
originates."

THE STATICS OF EVOLUTION

Muller was concerned not only with evolution as a process of change,
but also as a process maintaining the status quo. Muller's greatest
contribution to this, which he shares with Haldane (1937), is the

mutation load. They were the first to realize that the effect of mutation on fitness is independent of the effect on fitness of the individual mutations; the longer persistence of mild ones balances the greater effect of more drastic ones. In Muller's (1950) words each mutation leads to one "genetic death". Muller realized that the load was only half as large if the mutations were recessive, and could be further reduced if several could be picked off with one genetic death. This provided a way, although a rather abstract one, of assessing the total impact of mutation (Muller, 1950; Morton, Crow, and Muller, 1956). Although Muller realized that synergistic epistasis would work in this direction, he failed to realize the substantial load-reducing effect of rank order selection, even when truncation is inexact (Milkman, 1978; Crow and Kimura, 1979).

The insistence by Lerner and Dobzhansky that most loci consist of a large number of mutually heterotic alleles, struck Muller as being entirely without evidence and, moreover, contrary to genetic common sense. The result was a heated controversy (for a review, see Crow, 1987). It was finally settled, in Muller's favor, by careful experimentation in Drosophila and especially by new molecular techniques. I think the whole episode was unfortunate in at least one regard; it kept Muller — nearing the end of his active research life — from doing more important things.

THE EVOLUTIONARY ROLE OF SEXUAL REPRODUCTION

Muller shares with R. A. Fisher (1930) credit for the first convincing statement of the value of Mendelian segregation and recombination. It had been pointed out many times that the Mendelian shuffle produces novel gene combinations, thus helping the species find new ways of coping with an ever-changing environment. Recombination can sometimes put together new beneficial combinations, but it also takes them apart. So which process predominates?

Muller (1932a) argued that the advantage of sexual reproduction is putting together favorable mutations that arise in independent lines of descent. In an asexual system, two favorable mutations can be incorporated into a species only if the second occurs in a descendant of the individual in which the first occurred. In contrast, sexual mixing allows genes to be incorporated in parallel rather than in series. Muller correctly emphasized that the two mutations must be favorable individually as well as in combination; only in this case is recombination an advantage rather than a hindrance. This is important, for some evolutionists at the time believed that the advantage of Mendelism is to put together mutations that are deleterious individually, but which interact to produce a favorable result. No doubt such lock-and-key fittings are important in evolution, but they don't evolve readily under free recombination. If such interactions were the most important means of evolution, asexual reproduction would surely be the prevailing mode. Sewall Wright's shifting balance theory (1988 and earlier) is an attempt to get around this dilemma. Wright's theory can be described as a way of conferring some of the benefits of asexual reproduction on a sexual population.

Fisher had much the same ideas as Muller, but presented it in a quite different way. Muller developed this idea quantitatively (1932, 1958). He knew only very elementary mathematics — although he had wonderful intuitive insights, and would undoubtedly have been an excellent mathematician if he had had the training. It was easy to improve Muller's theory quantitatively and this was done (Crow and Kimura, 1965).

The Muller—Fisher idea is no longer regarded as the major reason for the evolution and maintenance of sexual reproduction, at least in multicellular eukaryotes, although it may have been at an early time in the history of life. Many writers have argued that the advantage lies in the capacity of a sexual population to respond to fluctuating environmental conditions. A Mendelian population has a large store of potential variability hidden beneath a much smaller overt variability. This can be useful in tracking an ever-changing environment. One might argue that a very high mutation rate would accomplish the same thing, but recombination is much superior in one regard. Recombination, by shuffling existing genes, recombines those that are to some extent pre-tested; the worst mutations have already been eliminated by natural selection. Since most environmental changes are somewhat gradual, the chances of producing combinations that would be favorable in a new environment are greatly enhanced if there is recombination. As far as I know, the first to write about this process with a specific model were Sturtevant and Mather (1938). There have been many variations on this theme in the ensuing years. A currently discussed theory is that of Hamilton et al. (1990), which notes that parasites constitute a particularly malevolent and diabolically unpredictable environmental component.

Another of Muller's ideas "Muller's ratchet" is currently widely discussed (Muller, 1964; Felsenstein, 1974; Maynard Smith and Nee, 1990). Muller noted that most individuals in a population carry one or more deleterious mutations, often only very slightly deleterious. In a small population there is a finite chance at any generation that there will be no mutation-free individual. The best individual in the population may have, say, one mutation, and there is no way to reproduce an individual with none (reverse and compensatory mutations seem quantitatively insufficient to help much). So the ratchet has turned one notch, that is, the best individual in the population has one mutation. Then the same thing can happen again, so that the best individual has two, and the ratchet has turned again, And so on. In a sexual population this is no problem, for the mutation-free type can be reconstituted by recombination between individuals having different mutations. The ratchet is more devastating, of course, in a small population or one that goes through size bottlenecks.

For a phenomenon as ubiquitous as sexual reproduction, one would like an explanation that applies equally to populations of all sizes. Another theory is one that Muller himself might well have thought of, given his emphasis on mutation as a generally harmful process. Getting rid of bad mutations efficiently is a necessary part of being able to exploit the minority of good ones. Kimura and Maruyama (1966) were the first to show mathematically that with some kinds of gene interaction sexual reproduction eliminates mutants from the population more efficiently that asexual. Several mutant genes could be eliminated at once with a single genetic death. The kind of epistasis required is that in which each additional mutation is worse than if the mutations acted independently.

This idea was not given major emphasis initially because of doubts about the epistasis — not the direction of epistasis, for there was plenty of empirical evidence that this was the prevailing type — but whether the degree of epistasis was quantitatively sufficient (Crow, 1970). It was also realized that truncation selection (saving only those individuals that are above a certain cut-off point) generated strong epistasis in the right direction. But there were serious doubts as to whether nature truncates. These doubts were removed when Milkman (1978) noted that even a crude approximation to truncation selection has almost the same statistical properties as strict truncation — an idea developed in more quantitative detail by Crow and Kimura (1979). If mutation rates are high, this becomes an important argument for sexual reproduction as a mechanism for reducing the mutation load. The strongest advocate of this view is Kondrashov (1988). The importance of this idea depends very much on a number that should be known in the near future, namely the total rate of occurrence of harmful mutations.

Alas, all this work was done after Muller's death. I mention it here, for I believe that Muller would be strongly supporting this idea were he still alive; it is fully consonant with his overall view of the important effect of mutation on fitness (Muller, 1950) and his realization that epistasis could reduce the mutation load. I don't know what he would think of Kimura's (1983) Neutral Theory, but he certainly would be conscious of the dilemma posed by 3 billion base pairs in the mammalian genome if all could produce deleterious mutations.

Along with the evolution of Mendelian recombination there has been a tendency for the most complicated organisms to be diploid. Why is diploidy advantageous? The most obvious reason, that it conceals deleterious recessives, is inadequate; for, despite an initial advantage after diploidization the population will soon reach a new equilibrium between mutation and selection. Furthermore, there are now twice as many genes to mutate so the mutation load is twice as large (unless the mutations are *completely* recessive, and few are). So, it may be that many species have evolved diploidy because of its immediate advantage and then were stuck with it; since going back to haploidy would have all the disadvantages of inbreeding (Crow and Kimura, 1965). Under somewhat restricted circumstances, however, (quasi-truncation selection, partial dominance, high mutation rate) diploidy can be advantageous (Kondrashov and Crow, 1991). The importance of this remains to be seen.

Another advantage of diploidy, and one that Muller advocated, is protection from somatic mutation (Crow and Kimura, 1965).[9] The somatic mutation clock is set back to zero each sexual generation, and diploid organisms are protected from the harmful effects of recessive (or partially recessive) mutations that occur in somatic cells — for example death of essential cells. Perhaps more important is protection from malignancy, for example childhood tumors such as retinoblastoma that require homozygous recessive mutations (natural selection is, of course, indifferent to post-reproductive deaths). Consistent with this view is that most species with elaborate and differentiated soma are diploid.

THE MENDELIAN BASIS OF SPECIES DIFFERENCES

Molecular studies have made this question seem anachronistic, but we must remember that in the 1930s there were still a number of biologists who argued that species and higher order differences were not chromosomal. Gene differences were important within a species, it was said, but not for determining differences between more distant groups. Since such groups do not hybridize, the question could not be definitively answered.

The work of Sturtevant and Dobzhansky (Dobzhansky, 1937) on hybrids between what were then called "races" of Drosophila pseudoobscura went part way toward dispelling such doubts. But it could be argued, and it was, that these were not "real" species. So Muller devised an ingenious way of bypassing the sterility of the hybrids between Drosophila melanogaster and D. simulans which, because of the complete sterility of their hybrids, were generally regarded as "good" species. Muller and his student Pontecorvo (1940) crossed triploid melanogaster females with heavily irradiated simulans males. The triploids produced a variety of chromosome combinations in their gametes, and the radiation produced nondisjunction of some of the simulans chromosomes. The result was a mixture of chromosomes from the two species, among which a few would happen to possess a full diploid complement. Detailed analyses of many such flies demonstrated the chromosomal basis for the sterility (and in many combinations, lethality) as well as various morphological differences in the two species. In particular a normal fly that had its major chromosomes entirely from melanogaster but had simulans cytoplasm was normally fertile. Differences between the species were shown to be chromosomal. I don't know whether this neat experiment was effective in convincing skeptics or not. A demonstration of genic similarities and differences between mice and elephants had to wait for the development of molecular methods.

ISOLATING MECHANISMS

In the 1930s there was a great deal of discussion of isolating mechanisms — various devices that keep species from crossing and thus having their differences swamped out by recombination. Dobzhansky's influential book, "Genetics and the Origin of Species," first published in 1937, gave a great impetus to research on speciation and isolating mechanisms. Like many present-day students of evolution, Muller regarded the origin of species as a rather small part of the large subject of evolution. The important question to him was how evolution progresses from simple to complex and ever-better adaptations, not the rather special problem of splitting into species. Species were but one stage, although a significant one, in a hierarchy leading to higher categories.

Nevertheless, Muller wrote several times about isolating mechanisms, for the subject was much discussed in the 1930s and early 1940s (e.g. Muller, 1938, 1942). Early on he argued that genes causing hybrid sterility or inviability were an inevitable consequence of the separate evolution of two groups that had been separated for other reasons, usually geographical. When two populations evolve independently they will evolve different ways of doing the same or similar things — especially, but not necessarily, if they find themselves in different environments. Sooner or later, interdependent sets of genes will evolve in each group — genes that work harmoniously with each other. When different sets of such genes are mixed, the harmony is upset. Muller cited an example from hybrids between *Drosophila melanogaster* and *D. simulans*; despite the identical bristle arrangements in the two species, the hybrids had an irregular pattern. The imbalance effect is especially great in recombinants, where various mixtures of the parental chromosomes are put together. Hence F_2 and backcrosses are usually less viable and fertile than F_1s. Yet such imbalance occurs even in the F_1 of the heterogametic sex, as Haldane (1922) had pointed out earlier. This theory was abundantly confirmed by many studies in several species showing that certain chromosome combinations are necessary for viability or fertility in inter-species recombinants.[10]

It might be thought that this emphasis on epistatic interactions is contradictory to Muller's emphasis on partial dominance and interlocus additivity in his other writings. But, as he pointed out, a large fraction of the genes can act roughly additively — indeed this must be the case or evolution in Mendelian populations would not work at all — and yet leave plenty of room for an important minority to have the kinds of interactions required for species isolation.

Muller always emphasized that species differences were a natural consequence of geographical separation leading to independent evolutionary processes, even when the environments appeared identical. He was pleased that his view of allopatric speciation, arrived at from genetic considerations, was in agreement with those of Ernst Mayr, arrived at from field and taxonomic observations. The subject is not discussed so much nowadays. The prevailing view is that allopatric speciation occurs and that Muller's explanation is sufficient; whether sympatric speciation is frequent is still not known. And of course, everyone agrees that, if the hybrids are inviable or sterile, there will be selection for avoidance of inter-species matings. In Fisher's words (1930, p. 130): "The grossest blunder in sexual preference, which we can conceive of an animal making, would be to mate with a species different from its own and with which the hybrids are either infertile or, through the mixture of instincts and other attributes appropriate to different courses of life, at so serious a disadvantage as to leave no descendants." That natural selection would lead to behavior that avoids this blunder is evident.

Muller gave a great deal of attention to chromosomal isolating mechanisms, particularly as they had been studied in Drosophila. His articles on this subject were done with typical Mullerian exhaustiveness, all feasible possibilities being considered in detail. For a richly detailed example, see Muller (1940).

DOSAGE COMPENSATION

Dosage compensation represents one of Muller's most strikingly original ideas. He, along with several others, had noticed that for most X-linked mutant genes the expression is the same in both sexes despite the fact that the male has only half as many X chromosomes (and the Y is essentially empty). For example, the eye color mutant "apricot" has the same eye color in males as in homozygous females. Yet females heterozygous for apricot and white eye have eyes only half as dark. This incomplete dominance is important, for if the apricot gene were completely dominant to white, there would be nothing to explain. The phenomenon showed up only with mutant genes, since the wild type (in this case red eye) gene is nearly completely dominant.

Muller envisaged a system of compensators, genes that work to equalize the gene effect in males and females. His mechanism has turned out be wrong. Rather, the compensation is brought about by greater transcriptional activity of X-linked genes in males (and in mammals, by the inactivation of all but one X chromosome in each cell). But the important point is that he saw that there was a problem, and with characteristic thoroughness and inventiveness set out to solve it (Muller, 1950a).

Now, what is particularly insightful is Muller's realization that there would be no need for dosage compensation if the normal alleles were fully dominant. Furthermore it is hard to imagine that a system of compensation would evolve solely to take care of rare mutations. Thus Muller reached the conclusion that the common wild type gene is not fully dominant, otherwise compensation would not be needed. Later studies, some initiated by Muller, have shown the correctness and generality of this observation. Almost always a detailed study of the heterozygote will reveal a detectable effect. Sometimes the effect is detected only statistically, as when the typical "recessive" lethal gene in nature reduces heterozygote viability by some 2—3 percent. Yet this small effect is enough to insure that the effective selection is on heterozygotes rather than on the much rarer homozygous mutant phenotypes.

This is a beautiful example of what was perhaps Muller's most striking intellectual trait: seeing far beyond the immediate and obvious consequences of an observation or experiment. From the observation that eye color *mutants* show the same phenotype in males and females, he reached the seemingly irrelevant conclusion that *normal* genes are only partially dominant. A remarkable piece of close reasoning.[11]

MULLER AS SELECTIONIST

Muller (1950) went further in his paper on dosage compensation to argue for what he called the "meticulousness of evolution." He emphasized that such fine-tuning, producing adjustments not directly observable and discerned only by highly indirect evidence, implies a "surpassing precision of adjustment". Here and in many other places, Muller emphasized the importance of genes with very minor effects, and therefore supported Darwinian gradualism. Muller not only was the first to realize the ubiquity of partial dominance, but he placed great importance on the fact that this meant that selection was correspondingly more effective — i.e., that heritability is higher.

Muller realized, of course, that genes often interact in complicated ways, but he thought that for the most part genes were selected on the basis of their additive properties. He didn't often use Fisher's vocabulary; if he had done so, he would have pointed to additive genetic variance for fitness as the rate-determining factor in evolution. He was aware of polymorphisms, but as I mentioned earlier, he was strongly opposed to the ideas being put forth by Lerner and Dobzhansky in the 1950s and 1960s which assumed ubiquitous overdominance. Muller did not live long enough to see his view upheld by molecular data.

Muller didn't care much for the idea, so strongly emphasized by Wright and others, of the selection-retarding effects of pleiotropy. In particular, he believed that pleiotropic effects need not be permanent. He said that if there is a gene that produces an improved eye and a worse wing, there will surely be modifiers that will alter only one aspect so that the eye and wing effect are not permanently locked together, and he pointed to examples of modifiers in Drosophila that change only one manifestation of a pleiotropic gene. Perhaps no leading evolutionary thinker, unless it be Fisher, had as much confidence in the efficiency and meticulousness of mass selection as Muller. This was one reason for Muller's confidence that eugenics could be effective.

Of course Muller was only one among a great many who emphasized Darwinian gradualism as opposed to the evolution by jumps then being advocated by several, most vocally by Goldschmidt (1940). Although Wright, Fisher, Haldane, and Muller had differences of emphasis, they were fully in agreement on this point. Undoubtedly these four, along with influential writers as Dobzhansky, Mayr, and Simpson, were mainly responsible for the nearly universal acceptance of the neo-Darwinian view, and the consequent demise of Goldschmidt's theory.

Those who follow the history of evolutionary thought will undoubtedly have plenty to debate in deciding who played the major roles in this widely held consensus. Surely Fisher's *Genetical Theory of Natural Selection* will be awarded an important place. Equally important at the time, although its lasting influence has been less, was Haldane's *The Causes of Evolution*. Shortly after came Dobzhansky's influential *Genetics and the Origin of Species*. Muller clearly added a strong voice and some of the cleverest kinds of evidence, and had a strong influence on those reading the Drosophila literature. But I don't know whether his emphasis on the larger evolution community was comparable to that of Fisher, Haldane, Wright, Dobzhansky, Mayr, and Simpson.

For one, such as myself, who was around at the time that Gold-schmidt had not yet been consigned to oblivion, the current discussion of "punctuated equilibrium" seems like reviving a very old issue. In 1947 Muller had this to say: "It is quite evident that the time-rate of change within individual lines and also the rate of diversification of lines has for some large groups been exceedingly different during some portions of geological history than during others; there have been unquestionable spurts and bursts, and contrasted long periods of stasis." Then he went on to argue against any view that required novel processes to account for periods of rapid change. A point that is not often mentioned in current discussions is that one reason for Gold-schmidt's views not carrying much weight was that, in distinction to Muller, Sturtevant, and Wright, some of his experimental work did not stand up, making people less accepting of his speculations. He got off on the wrong foot with the Morgan school by arguing that linkage maps were a meaningless artifact (Goldschmidt, 1916).

MULLER AND HUMAN EVOLUTION

Muller often wrote about human evolution, but most of what he wrote on this topic does not stand out as strikingly different from what others said.

The idea of kin selection comes from Fisher (1930, p. 159) and Haldane (1932, p. 131), and was developed in detail by Hamilton and others. Muller clearly had the same idea, perhaps independently (e.g., Muller, 1967). He, like Haldane, attributed much of what is noble and unselfish about our own species to the tribal structure of our recent ancestors. Since everyone in a small tribe is likely to be related to everyone else, this could lead to the development of cooperative and altruistic behavior because of shared genes. Muller did not use the vocabulary, but this is standard kin-selection (although based on group membership rather than kin recognition). Then, these good habits became reinforced by social pressures as civilization became more advanced. Thus, it is "human nature" to be friendly to members of one's own group. It is equally "human nature" to be unfriendly to members of other groups. Muller believed that we would need the mollifying effects of a developing civilization, and perhaps of eugenics, to expand the former tendency and reduce the latter.

WHAT OF THE FUTURE?

No discussion of Muller's views of evolution would be complete without mention of his view of future human evolution. From his student days, Muller was an ardent advocate of eugenics. Yet he was highly critical of the coercion, the genetic naiveté, and simplistic assumptions of the early American eugenics movement. He argued that in a society with such environmental inequalities as existed in the United States, any kind of fair genetic assessment was totally unrealistic.[12] His left-wing views led him to go to the Soviet Union in the 1930s. One of his hopes was that social inequities would be minimized there, making eugenic assessments feasible. His discouragement with American eugenics is clear in a paper he gave at the 1932 Eugenics Congress, entitled "The dominance of economics over eugenics"; it turned out to be a mortal blow to the already weakened American eugenics movement (Muller, 1933).

Nevertheless, Muller remained an ardent eugenicist. His objections to the earlier movement were its simplistic genetic assumptions, its association with racism and class differences, its advocacy of compulsion, and its emphasis on negative aspects. Muller was never content that the major aim of eugenics should be that of ridding the population of severe genetic disease. He hoped for a human species that was far better than ourselves. He referred to current *Homo sapiens* as "hastily made-over apes", and asked how we would like it if our Neanderthal ancestors had been able to halt further evolution. He would like to be a member of a population in which the average person could have the joys of an easy understanding of relativity and quantum electrodynamics, of the complexities of a Bartok or late Beethoven Quartet, or of the poetry of Rilke (my words, his ideas). He saw a future population with cheerful, cooperative disposition, high intelligence, artistic appreciation, and good health for almost everyone. He advocated artificial insemination as the principal means to this end, and no doubt were he still alive he would urge supplementing this with the latest molecular technology. He argued against using sperm from a living person. By storing semen until after death, one could not only select against diseases with late onset (which natural selection had ignored), but would diminish the possibilities of self-aggrandizement on the part of sperm donors. He wanted women to make the choice. He was adamant in his rejection of any compulsion or artificial incentive.

In Muller's (1967) words, "It seems unthinkable that if man brought under his control the evolution of the major species about him he would wish to leave his own nature to the play of uncontrolled forces." I don't need to say that Muller's program was greeted with opposition and, I suspect even more annoying to him, by indifference. In Muller's view most of his contemporaries were Luddites. Few shared his view that people would behave as rationally as he expected. A paradox in Muller's life was his frequent mistrust of people individually along with his confidence in rational human behavior in the larger society.

WHAT IS MULLER'S POSITION AMONG EVOLUTIONISTS?

I think the time is not yet ripe to have a proper assessment, and in any case it needs to be done by a much more thorough and scholarly study than I have essayed. How does one compare Fisher, Wright, Haldane, and Muller?[13] With reservations, here is a provisional assessment of the four.

For sheer breadth of knowledge, Haldane was far ahead. He was blessed with a memory that was truly remarkable. He could (and would, on slight provocation) recite large sections from Shakespeare, Dante, the Koran, or Hindu mystics. He was fluent in Latin and Greek at an early age.[14] He rarely had to look anything up and it is said that he trusted his memory enough not to refer to earlier pages of his mathematical notes. He must have worked at breakneck speed in order to have done all that he did.[15] Genetics was only part of his writing. He also wrote knowledgeably on astronomy, enzymology, physiology, life under several atmospheres of pressure, statistics, and mathematics, among many other topics.

Haldane's *Causes of Evolution* was well received at the time of publication (1932) — more favorably than Fisher's *Genetical Theory of Natural Selection*.[16] Yet, although it is an excellent account of evolution — good, provocative reading still — it lacks Fisher's mathematical creativity. Haldane's mathematical studies were characterized by brute force rather than finesse. His most widely recognized mathematical contributions are the ideas of genetic load, which Muller (1950) arrived at independently, and the cost of evolution; both were highly influential, but mathematically elementary. Haldane's unique contribution was to use his great erudition to do research and make substantial advances in highly diverse areas, which add up to a great deal. He isn't remembered for any special thing (an exception might be his work on enzyme kinetics), and he didn't found a school; his interests were too broad and he was too open-minded. And we mustn't forget his wide influence as a popularizer.

Wright was like Haldane in one respect. Both of them, when confronted with a problem, would simply grind out the mathematics wherever it led. Neither looked for more elegant ways; neither had Fisher's mathematical skill and inventiveness, although Wright managed to solve some very difficult problems by his own rough and ready methods. They differed in that, whereas Haldane looked at every problem under the sun, Wright repeatedly came back to one idea — his shifting balance theory. He is remarkable for having had the idea at a very early age, in the 1920s, and continuing to espouse it the rest of his life. Although he added details and improved the presentation, he hardly changed the basic idea at all. Wright's permanent place in evolutionary history will depend very much on whether the shifting balance turns out to be important. The jury is still out. I should add that much of Wright's life, surely the great majority until he moved to Wisconsin, was devoted to his experimental studies of guinea pigs. Evolution, although his deepest interest, did not make the major claim on his time and energy. Wright, like Fisher and Haldane, contributed greatly to the methodology of population genetics, especially in the areas of inbreeding and population structure.

Of the three, Fisher was far and away the most skilled and inventive mathematically. He solved very difficult problems, and almost always with elegance and grace. Often, he invented a totally new method. I suspect that, whereas the mathematical methods of Haldane and Wright will be superseded (many already have been), many of the innovations introduced by Fisher will be permanent. With each rereading of *The Genetical Theory of Natural Selection* I find something new, further evidence of his astonishing creative insight. So I suspect that this book will be remembered along with Darwin; one could make the case that it is the most important book on evolution since Darwin. Whereas Haldane and Wright impress one as having powerful minds, Fisher seemed to me to have a special, qualitatively different creativity, especially when mathematics was involved and still more especially when a geometric formulation was possible.

How does Muller compare with this Holy Trinity? If a theory of evolution is to be mathematical, he doesn't stack up very well. Nevertheless, Muller's insight and power of thinking were such that he often arrived at solutions in his own ways that would ordinarily have been reached mathematically.[17]

Like Haldane, Muller wrote on many different evolutionary subjects. He and Fisher both displayed a striking originality, Fisher in mathematics and Muller in biology. Muller's contributions were mainly qualitative, not quantitative. His ideas on the gene as the basis of life, dosage compensation and partial dominance, the meticulousness of selection, the nature of mutation as the ultimate evolutionary process, the importance of sexual reproduction, and the Mendelian nature of species differences, to say nothing of his experimental work, surely entitle him to a place along with the big three.

I sometimes wish that he had written a book at the same time that Fisher and Haldane did theirs. He might have produced something comparable to Fisher's block buster. It would not have had Fisher's mathematical originality, but it would have had Muller's own kind of creativity.

Wright wrote little about human evolution. Fisher was an active eugenicist in his earlier days, and thought he had perceived the major cause for the decay of civilizations. Haldane was an active communist propagandist until his break with the party, and often commented on political questions. Muller, alone among the three, was a genetic crusader throughout his life. One cause was the dangers of mutation, especially radiation-induced. The other was eugenics. He had an unbounded faith that mankind could and would better itself. He rejected the "stultifying assumption" that people would have to be coerced rather than inspired to engage in positive eugenics. Continuing from the article that produced my introductory quote:

Who can say how far this seed of self-awareness and self-transfiguration that is within us may in ages to come extend itself down the corridors of the cosmos, challenging in its progression those insensate forces and masses in relation to which it has seemed to be but a trivial infestation or rust?[18] For the law of the gene is ever to increase, and to evolve to such forms as will more effectively manipulate and control materials outside itself so as to safeguard and promote its own increase. And if the mindless gene has thereby generated mind and foresight and then advanced this product from the individual to the social mind, to what reaches may not we and our heirs, the incarnations of that social mind, be able, if we will, to carry consciously the conquests of life?

It will be some time before we know whether Muller was a prophet, ahead of his time, or whether mankind will never make the attempt to evolve to a higher plane.

CONCLUSION

For his numerous contributions to evolution — the gene as the basis of life, the origin of new genes by duplication of old ones, the nature of the mutation process, the change of mutation rates by radiation, the evolutionary advantages of recombination and diploidy, the necessity for dosage compensation, the ubiquity and evolutionary significance of partial dominance, the mutation load, the evolution of isolating mechanisms, and the chromosomal basis of species. differences, to say nothing of his multitude of experimental contributions to genetics — entitle Muller to recognition as a leader in the development of the neo-Darwinian view of evolution.[19]

University of Wisconsin

NOTES

[1] Wright, modest as always, often mentioned Muller's greatness. The University of Texas was fortunate, he said, that he turned down a job later given to Muller. Haldane (1947), not inhibited by false modesty, was more quantitative. In the course of disagreeing with Muller's eugenic proposals he said: "Dr. H. J. Muller has suggested a method for the radical improvement of the human race, involving the widespread use of artificial insemination. I guess that if I were made eugenic world dictator I should have one chance in a hundred of choosing the right path. Dr. Muller is ten times as good a geneticist as I, so he might have one chance in ten, but not, I think, much more."

[2] Muller wrote some 350 articles, many of which are reprinted (usually only in part) in his collected papers (Muller, 1962). A full length biography, with both scientific and personal information, was written by his student Carlson (1981).

[3] I think it is significant that Muller, whose busy life permitted little leisure, nevertheless stole time to read science fiction. His speculations of possible human futures included many far-out ideas. Muller was not facile in speech and wondered about the future possibility of direct transfer of thought, bypassing the slow and clumsy processes of speaking and writing.

[4] It is a minor tragedy in the history of genetics that Morgan began his studies with

Drosophila melanogaster. He could as easily have started with *D. simulans*, which has the great advantage for evolutionary study of producing fertile hybrids with several other species, which *D. melanogaster* does not.

[5] While writing the one paper on which Muller and I collaborated (Morton *et al.*, 1956) he argued that we should include a point that I thought was tangential. He feared that if we didn't include it people would assume that we hadn't thought of it. Incidentally, my view prevailed; despite a reputation to the contrary, Muller did not always insist on having his way.

[6] Here is an example from these files, a letter to Muller's secretary from a New York hotel: "We are very sorry to advise you that after a thorough checking of room 1201A, which Dr. Muller occupied during his recent stay with us, no rubbers were found."

[7] Muller, himself, was to be the first to demonstrate the falsity of this by his experiments showing the mutagenic effects of ionizing radiation (Muller, 1927). Earlier he had shown the dependence of the mutation rate on temperature and noted that this argued for mutation being a chemical process.

[8] Carlson (1981) records that Henry Fairfield Osborn, who had given the welcoming address at the Congress, thought this was a joke and congratulated Muller on his sense of humor.

[9] This offers me the chance to give an example of Muller's obsession with recording his ideas. This idea occurred to me and I proudly told Muller about it. He readily agreed, but pointed out that he had actually thought of it before. I wasn't surprised, knowing Muller's capacity to think of almost everything, but I *was* surprised when he went to considerable pains to dig up some old notes documenting the idea, apparently feeling that he had to convince me with written proof.

[10] Muller (1940) preferred Haldane's (1922) original explanation, the upset of X-autosome balance in heterogametic hybrids, to his later (1932) explanation based on upset X-Y interactions.

[11] I once told an audience that I envied them the pleasure that I could no longer have, that of reading Muller's paper on dosage compensation for the first time.

[12] He differed strikingly from Fisher in this regard. Fisher thought existing differences in social standing were positively correlated with genetic potential.

[13] I count myself fortunate in having known these great figures in the development of the neo-Darwinian theory. I knew Haldane casually, Fisher quite well, Muller very well, and Wright from almost daily association for more than 30 years. Nevertheless, in passing judgment on such giants, I feel a bit like a dachshund comparing the heights of giraffes.

[14] My late colleague, Klaus Pätau, enjoyed reciting Virgil in the original Latin. During one party he had a memory lapse and Haldane finished the poem for him.

[15] A consequence of his rushing through algebra and computations and of the fact that his memory was only *almost* perfect, his papers have a number of troublesome minor errors.

[16] The 1990 edition of Haldane's book includes a perceptive foreword and afterword by Egbert Leigh that discuss Haldane's work in terms of later developments.

[17] I observed this many times. Muller was once a "student" in my population genetics course, and several times provided an intuitive, verbal way to arrive at a conclusion that I had presented mathematically.

[18] The earth is much more highly oxidized than before life began. Could Muller have intended a double meaning?
[19] This is contribution number 3186 from the Laboratory of Genetics.

REFERENCES

Carlson, E. A. (1981), *Genes, Radiation, and Society. The Life and Work of H. J. Muller*, Cornell University Press, Ithaca, NY.

Crow, J. F. (1970), 'Genetic Loads and the Cost of Natural Selection', in K. I. Kojima (ed.), *Mathematical Topics in Population Genetics*, Springer-Verlag, Berlin, pp. 128—177.

Crow, J. F. (1987), 'Muller, Dobzhansky, and Overdominance', *Journal of the History of Biology* 20: 351—380.

Crow, J. F. and M. Kimura (1965), 'Evolution in Sexual and Asexual Populations', *American Naturalist* 99: 439—450.

Crow, J. F. and M. Kimura (1979), 'Efficiency of Truncation Selection', *Proceedings of the National Academy of Sciences USA* 76: 396—399.

Dobzhansky, Th. (1937), *Genetics and the Origin of Species*, Columbia University Press, New York.

Felsenstein, J. (1974), 'The Evolutionary Advantage of Recombination', *Genetics* 78: 737—756.

Fisher, R. A. (1930), *The Genetical Theory of Natural Selection*, The Clarendon Press, Oxford.

Gilbert, W. (1986), 'The RNA World', *Nature* 319: 618.

Goldschmidt, R. (1916), 'Crossing over ohne "Chiasmatypie"?', *Genetics* 2: 82—95.

Goldschmidt, R. (1940), *The Material Basis of Evolution*, Yale University Press, New Haven.

Haldane, J. B. S. (1922), 'Sex-Ratio and Unisexual Sterility in Hybrid Animals', *Journal of Genetics* 12: 101—109.

Haldane, J. B. S. (1932), *The Causes of Evolution*, Harper, New York. Reprinted 1990 with a Foreword and Afterword by Egbert Leigh, Princeton University Press, Princeton, N.J.

Haldane, J. B. S. (1937), 'The Effect of Variation on Fitness', *American Naturalist* 71: 337—349.

Haldane, J. B. S. (1945), 'A New Theory of the Past", *American Scientist* 33: 129—145, 188.

Haldane, J. B. S. (1947), 'Human Evolution: Past and Future', in G. L. Jepsen, E. Mayr, and G. G. Simpson (eds.), *Genetics, Paleontology, and Evolution*, Princeton University Press, Princeton, New Jersey, pp. 405—418.

Hamilton, W. D., R. Axelrod, and T. Tanese (1990), 'Sexual Reproduction as an Adaptation to Resist Parasites', *Proceedings of the National Academy of Science U. S. A.* 87: 3566—3573.

Kimura, M. (1983), *The Neutral Theory of Molecular Evolution*, Cambridge University Press, Cambridge.

Kimura, M. and T. Maruyama (1966), 'The Mutation Load with Epistatic Gene Interactions in Fitness', *Genetics* **54**: 1337—1351.

Kondrashov, A. (1988), 'Deleterious Mutations and the Evolution of Sexual Reproduction', *Nature* **336**: 435—440.

Kondrashov, A. and J. F. Crow (1991), 'Haploidy or Diploidy: Which is Better?', *Nature* **351**: 314—315.

Maynard Smith, J. and S. Nee (1990), 'Clicking into Decline?', *Nature* **348**: 391—392.

Michod, R. E. and B. R. Levin (eds.) (1988), *The Evolution of Sex*, Sinauer Associates, Sunderland, Mass.

Milkman, R. (1978), 'Selection Differentials and Selection Coefficients", *Genetics* **88**: 391—403.

Morton, N. E., J. F. Crow, and H. J. Muller (1956), 'An Estimate of the Mutational Damage in Man from Data on Consanguineous Marriages', *Proceedings of the National Academy of Sciences U.S.A.* **42**: 853—863.

Muller, H. J. (1922), 'Variation Due to Change in the Individual Gene', *American Naturalist* **56**, 32—50.

Muller, H. J. (1927), 'Artificial Transmutation of the Gene', *Science* **66**: 84—87.

Muller, H. J. (1929), 'The Gene as the Basis of Life', *Proceedings of the International Congress of Plant Sciences* **1**: 897—921.

Muller, H. J. (1932), 'Further Studies on the Nature and Causes of Gene Mutations', *Proceedings of the Sixth International Congress of Genetics* **1**: 213—255.

Muller, H. J. (1932a), 'Some Genetic Aspects of Sex', *American Naturalist* **68**: 118—138.

Muller, H. J. (1933), 'The Dominance of Economics over Eugenics', *Scientific Monthly* **37**: 40—47.

Muller, H. J. (1936), 'Bar Duplication', *Science* **83**: 528—530.

Muller, H. J. (1939), 'Reversibility in Evolution Considered from the Standpoint of Genetics', *Biological Reviews* **14**: 261— 280.

Muller, H. J. (1940), 'Bearings of the Drosophila Work on Systematics', in J. Huxley (ed.), *The New Systematics*, The Clarendon Press, Oxford, pp. 185—268.

Muller, H. J. (1942), 'Isolating Mechanisma, Evolution, and Temperature', *Biological Symposia* **6**: 71—125.

Muller, H. J. (1947), 'Redintegration of the Symposium on Genetics, Paleontology, and Evolution', in *Genetics, Paleontology, and Evolution*, Princeton University Press, Princeton, N.J., pp. 421—445.

Muller, H. J. (1947a), 'The Gene', *Proceedings of the Royal Society B* **134**: 1—37.

Muller, H. J. (1950), 'Our Load of Mutations', *American Journal of Human Genetics* **2**: 111—176.

Muller, H. J. (1950a), 'Evidence of the Precision of Genetic Adaptation', *The Harvey Lectures* **18**: 165—229.

Muller, H. J. (1955), 'Life', *Science* **121**: 1—9.

Muller, H. J. (1958), 'Evolution by Mutation', *Bulletin of the American Mathematical Society* **64**: 137—160.

Muller, H. J. (1962), *Studies in Genetics: The Selected Papers of H. J. Muller*, Indiana University Press, Bloominton.

Muller, H. J. (1964), 'The Relation of Recombination to Mutational Advance', *Mutation Research* **1**: 2—9.

Muller, H. J. (1967), 'What Genetic Course Will Man Steer?', *Proceedings of the Third International Congress of Human Genetics*, pp. 521—543.

Muller, H. J. and G. Pontecorvo (1940), 'Recombinants between Drosophila Species the F1 Hybrids of Which Are Sterile', *Nature* **146**: 199—200.

Sturtevant, A. H. and K. Mather (1938), 'The Interrelations of Inversions, Heterosis and Recombination', *American Naturalist* **72**: 447—452.

Wright, S. (1988), 'Surfaces of Selective Value Revisited', *American Naturalist* **131**: 115—123.

WILLIAM C. WIMSATT

GOLDEN GENERALITIES AND CO-OPTED ANOMALIES: HALDANE VS. MULLER AND THE DROSOPHILA GROUP ON THE THEORY AND PRACTICE OF LINKAGE MAPPING

I. INTRODUCTION — A MYSTERIOUS CONFLICT

Such an attempt to find a general formula is futile, for it has been proved that the relation between map-distance and observed percentage of crossing over is different not only in different chromosomes, but even very strikingly in different regions of the same chromosome. The only satisfactory representation of such relationships is one that gives for each chromosome, and for each region of that chromosome the observed relation between map-distance and crossover values. From such careful, detailed region-by-region studies, a more general statement and formulation should emerge. Until such studies are completed, there is nothing to be gained from *a priori* attempts to formulate the relationship. . . . it is evident from the foregoing that within a given region the exact function of distance represented by crossover values depends directly upon the magnitude of the interference acting at each given distance. . . . (*MMH*, 1922, p. 171.)

Thus the authors of the *Mechanisms of Mendelian Heredity* — the second edition of their textbook manifesto — concluded their discussions of Haldane's views on a theory of linkage. By 1922, Drosophila genetics had center stage in theoretical investigations of the nature of the gene and their organization into chromosomes, whose mechanics gave what was soon to be known as "classical" genetics a theoretical centrality and predictive power rivalling that of classical mechanics in its own domain two centuries before. It appeared to be a picture of great generality. The key to this theoretical revolution in genetics was the discovery, analysis, explanation, and exploitation of linkage, and the linkage map was, properly, an exemplar of high theoretical science at its best. As authors of such a theoretical achievement, Morgan, Sturtevant, Bridges, and Muller — so a naive view goes — should have been both prepared and inclined to appreciate further theoretical developments of their model.

Haldane's views had been expressed in an elegant paper of 1919 in which he gave a general and systematic mathematical treatment of linkage — one which, one might have thought, would have been

107

Sahotra Sarkar (ed.), The Founders of Evolutionary Genetics, 107—166.
© 1992 *Kluwer Academic Publishers. Printed in the Netherlands.*

welcomed as friendly theoretical support. But surprizingly, the Morgan school had very little use for Haldane's abstract mathematical models, general though they might be. Indeed, that was perhaps their main problem with his approach, for it seemed to them to paper over the variability and complex physical interactions which they felt were central to any account of recombination. And yet the map itself was based on resultant recombination frequencies — whatever their causes — and not on actual physical distances along the chromosome. It might thus already seem to have taken these complexities into account, and other complexities, including interference, were directly addressed by Haldane in his paper. How did their differences with Haldane manifest themselves in their views and in their methodology? Was the argument about an *index* of interference — their "coincidence" measure — which they used and the English did not? (The next passage was an extended advertizement for their measure.) Was it just a tiff along nationalistic lines — perhaps mediated by further transatlantic misunderstandings? (Haldane's paper sometimes seemed to show an almost obstinate failure to understand their claims, and to "correct" them by offering as an "alternative" the views they actually held.) Did the data compel an answer? (They certainly thought so, but many of the relevant questions — particularly over how much generalization is too much — are as much debated today as they were then.) Some of these issues go beyond the reasonable scope of this paper, and will be addressed at more length on another occasion. Here, I will compare the approaches of H. J. Muller — theoretician of the Drosophila group and J. B. S. Haldane to problems of linkage and recombination. They are indicative of the problems theorists and experimentalists have in talking to each other everywhere, and I suspect that many or most of the reasons are generalizeable.

Before embarking on this comparison of Muller and Haldane, it is appropriate to set the stage with what was known as of the fall of 1915, and immediately preceding, when Muller did his experiments and wrote up his dissertation work.

II. SOME BACKGROUND (AND SLIGHTLY WHIGGISH) HISTORY AND A PUZZLE

A somewhat standard account of the development of the linear linkage model of the localization of the genes on chromosomes proceeds something like this. (Wimsatt, 1987): The model of the Morgan school has a natural ancestor in the Sutton-Boveri hypothesis of 1902-3 that the genes are identical with, or located on or in, the chromosomes. Few biologists rushed to embrace this powerful hypothesis immediately however, largely because it has nothing to say about development or (what we now call) gene action, and it was widely supposed that any adequate theory of heredity would explain development as well as the hereditary resemblance of organisms. The Boveri-Sutton model itself seems to be passively correlational with respect to the activities of the genes, in which all of the interesting properties of the factors arise because they are associated (and therefore move with) the chromosomes.[1] For this and other reasons, Morgan himself is a sceptic not only of the Sutton-Boveri hypothesis, seeing it as "preformationist", and "deterministic" but even of Mendelism, which he fears can be used to explain virtually any patterns of inherittance whatsoever if there are no constraints on the positing of additional factors to explain the results (Morgan, 1909, 1910a). (For somewhat different but mostly complementary accounts of Morgan's conversion see Allen, 1979, and Lederman, 1989). Nonetheless, only a year later, Morgan finds unusual patterns of inheritance of a white eye mutant in *Drosophila*, and to explain them, advocates an account according to which the factor for the white eye mutant is at first thought to be carried with (1910b), and finally located on (1911a) the X-chromosome. (Moore, 1972, has a detailed comparison of these hypotheses).

In science, hypotheses which appear to work are rapidly employed as tools to bootstrap other results. Using a strange "criss-cross" pattern of inheritance characteristic of the white-eye mutant as a diagnostic tool (see, e.g., Moore's (1972) discussion of sex-linkage), Morgan and his students quickly find other mutants which show this pattern, and which they (thus, by hypothesis) also locate on the X-chromosome (Morgan, 1911b). If the factors are located on the same chromosome and the chromosome is never broken up, one would expect these factors always to be inherited together — either both would be present or neither — a phenomenon predicted by Sutton in 1903. If, by contrast, the factors were on different chromosomes, and those chromosome assort independently, one would expect no statistical association of the characters produced by these factors in the offspring, as with Mendel's "second law" of independent assortment.

Anomalously, a pattern of association midway between these two extremes was found, with a constant characteristic frequency of association between the traits caused by factors or their alleles, greater than that expected with random or independent assortment, but less than universally correlated. Moreover, this constant and characteristic frequency varied with the pair of factors considered. Cytological work of Janssens (1909) showed homologous chromosomes intertwined around one another at an early stage of meiosis. This led Morgan to suggest that the genes are linearly arranged along the chromosomes, and further, that the tangling noted by Janssens leads them to break and interchange segments when they are separated in the cell cycle, producing separation of the factors in the gametes, and ultimately in the offspring. This would produce the observed pattern of association intermediate between those expected on Mendel's first and second law. On this hypothesis, he further suggests that the frequency with which they stay together or separate (which is constant for any given pair of characters, but differs for different pairs) should be a product of the frequency of breakage between the chromosome loci containing the factors responsible for these characters. This in turn should be a function of the distance between the factors along the chromosome (1911b).

Building on this, Sturtevant in 1913 shows how crossover frequencies can be used to generate the relative order of the factors or genes on the chromosome in what later came to be called a "linkage map". From the first published work, (the idea actually first came to him in 1911, as an undergraduate, so he had considerable time to digest and discuss it) Sturtevant shows a sophisticated conception of this theoretical structure, and the complex physical interactions through which it is tied to the material chromosome. Thus he resisted straightforward connections between his "linkage map" and the physical chromosome, pointing out that relative distances in the map do not imply relative distances along the chromosome because different parts of the chromosome may differ in their resistance to breakage.[2] Nonetheless, he argues that this conceptual device can be used to make new predictions about the recombination frequencies expected for new pairs of factors which can be located relative to each other in the map.

Sturtevant also notes that factors which are relatively close together should be used in constructing the map, for they and they only will have additive or nearly additive recombination frequencies or map distances. Factors which are further apart may have two or more crossovers occurring between them. These double crossovers will look like no crossovers — they will not be detected, since the two factors being tracked will end up on the same side, and in the same chromosome, as if nothing at all had happened. Double and higher-order crossovers will therefore give a biased measure — an underestimate of the distance between the factors being mapped. Finally, he also notes that if a single crossover occurs, it appears to make a second crossover less likely nearby than had that first crossover not occurred — a process that Muller later names "interference".

With this, Sturtevant has basically given all of the conceptual apparatus needed for a phenomenological theory of linkage — (1) a linear map constructed using (the smaller, but not too small) recombination frequencies among nearest neighbor factors which are additive or nearly so, as measures of map distance, allowing a representation of the order of the genes in the chromosome; (2) a recognition that more than one crossover between factors was possible, and that double crossovers between the closest factors being followed in a cross would "cancel" and appear as no crossovers, thus leading to underestimates of the recombination frequency or map distance; (3) that the frequency of double crossovers between factors should increase with increased separation between the factors, thus introducing a systematic "shortening" bias which increasingly affected estimates involving more distant factors; and (4) a recognition that crossovers appear to inhibit other nearby crossovers; (5) A judgement that this inhibition occurs presupposes a model or assumption for how frequent double crossovers should be if there is no mutual inhibition or "interference", and this last basically gives the model which leads to what we call the Haldane Mapping Function, though as we shall see, it could almost equally well be called the Trow model, or even the Muller or the Sturtevant model. Members of the Morgan school are quite sophisticated about the relation between the causal factors affecting recombination, and so is Haldane, though sophisticated, as we shall see in quite different ways.

So, the Whig history continues, with the Drosophila group discovering mutations, elaborating their maps of the X and other (autosomal) chromosomes, discovering other interesting phenomena (especially non-disjunction) which further supported their model, responding to alternative models and usually misdirected criticism put forward by Castle, Bateson and Punnett, Goldschmidt, and others, and generally winning the high ground as the established model by about 1920. (For reviews of this period, see Carlson, 1967, Allen, 1979, Carlson, 1981, Wimsatt, 1987, and Darden, 1991.) In 1919 Haldane writes his paper on linkage, which elegantly derives what is now called the Haldane mapping function, tidying up and formalizing what the Morgan school must have known all along. But this is getting ahead of the game. There is one more thing to do before we go to Muller's work, and then to Haldane's. This is to discuss the origin and understandings of Trow's formula, since it plays an important role in the understanding both Muller and Haldane have of the explanations of linkage. This takes going a little out of order, since it requires some reference to Haldane's 1919 paper.

III. TROW'S FORMULA

To my surprize, what we know as "Haldane's mapping function" is inspired by a formula for calculating recombination frequency given not originally by Haldane, nor by any of the Drosophila group, but attributed by Haldane to Trow (1913).[3] As such it was commonly regarded as a consequence of a theory (the reduplication theory of Bateson and Punnett (1911)) which comes under increasing attack throughout the period, and probably persists for as long as it does (more in England than in the United States) because of the enormous influence of William Bateson, who is one of the last holdouts from accepting the views of the Morgan school. Nor did it represent Haldane's best attempt (in that paper or since — Haldane (1931)) to model the phenomena of linkage. We will return to this, but for the present, Trow's formula did provide a common basis of experience for all concerned.

In fact, the proper attribution of this formula is somewhat problematic. Muller (1916—II pp. 285—8, esp. p. 286) explicitly claims that they had it before Trow published, but didn't publish it (the data or the formula?) because the data did not fit the formula. Sturtevant's discussion in his 1913 paper is consistent with the use of Trow's formula, though he actually argues for interference not by showing deviation from Trow's formula but, more empirically, by showing that from his data, there are fewer crossovers in a given region in the presence of a crossover elsewhere in the chromosome than in its absence. So unlike Trow, the members of the Morgan school do not believe that the formula is true — honoring it in the breach, so to speak, by pointing out

that it does not fit their data. In 1914, Sturtevant devotes a whole paper
to showing why not, in the course of which (pp. 542—3) he reformu-
lates "Trow's formula" so as to give gametic ratios directly, in the form
later used by Haldane. He particularly acknowledges Muller's help in
the 1913 paper, and Muller's help is likely also for the 1914 paper.
This latter paper is cited by Haldane, who may refer to the formula as
"Trow's formula" because Sturtevant does so. See also Muller, 1916—
II, pp. 285—288 for more discussion of "Trow's formula" and a
penetrating critique of the methodology of the advocates of the
reduplication hypothesis.

One of the remarkable things about Trow's formula is that it originated in a competing theory, yet appears to capture the effects of
double crossing over for the linear linkage model. This competing
hypothesis was the reduplication hypothesis of Bateson and Punnett
(1911), which purported to explain the non-random frequencies of
association of the factors in the gametes in terms of differential division
rates of cells containing different combinations of factors. (See the
diagram on how reduplication is supposed to produce these ratios on p.
317 of Trow, 1913 — reprinted as Figure 1. Sturtevant's 1914 paper
cited above is in fact a critique of the reduplication hypothesis.) In
attempting to explain the phenomena of linkage in developmental
terms, the reduplication hypothesis resonated with sceptics who saw
chromosomal mechanics as too performationist or too mechanical and
as seeming to ignore cellular interactions in development. (Compare
Morgan, 1909, before his conversion to the chromosome theory! See
also Coleman, 1970.)

Trow's formula works also for the effects of multiple crossovers
because both theories supposed that the crossovers or reduplications
which produced the systematic distortions in frequencies were probabilistically independent events. To a first approximation (i.e., ignoring
the fact that it doesn't actually fit the data of the Drosophila group!),
Trow's formula should appear (to philosophers) to be a nice example of
the underdetermination of theory by the data. It is likely the association
of what we now know as "Haldane's mapping function" with "Trow's
formula" did not incline the Drosophila group to use it — indeed quite
the reverse. Not only did they see it as inaccurate (because of the
effects of interference — see Sturtevant, 1914, p. 548), but at least in
England — due no doubt to Bateson's enormous influence — the
reduplication theory was still a live if increasingly moribund competitor

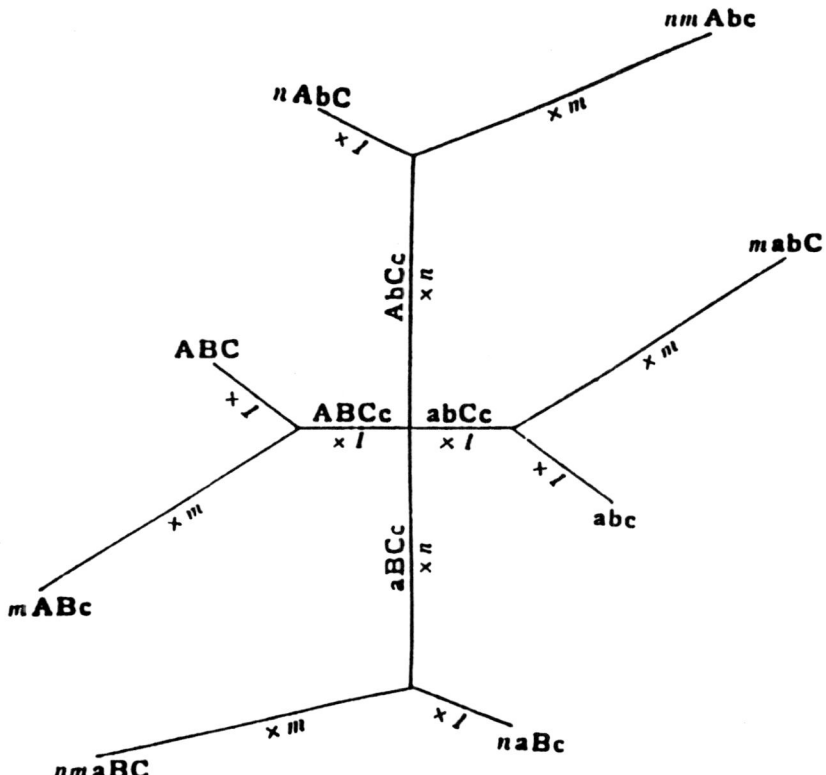

Fig. 1. Trow's diagram of lineages of cell division illustrating how the ratios of gametes containing different combinations of factors were supposedly produced, producing the phenomena of "linkage' according to the reduplication hypothesis of Bateson and Punnett. "Trow's Law" can be read off directly from these ratios. From Trow, 1913, p. 317. Reproduced through courtesy of *Journal of Heredity*.

as late as 1919 or 1920, and its legacy survived even later — even down to today in use of the terms "coupling" and "repulsion" to describe *cis* and *trans* forms of double heterozygotes. (Thus Sinnott and Dunn (1932, p. 147) set it as a problem for students to compare the two theories, and argue for which is better and why.) Without doing a systematic survey, I estimate that more papers were written discussing or using it than any of the competing theories that the Morgan school

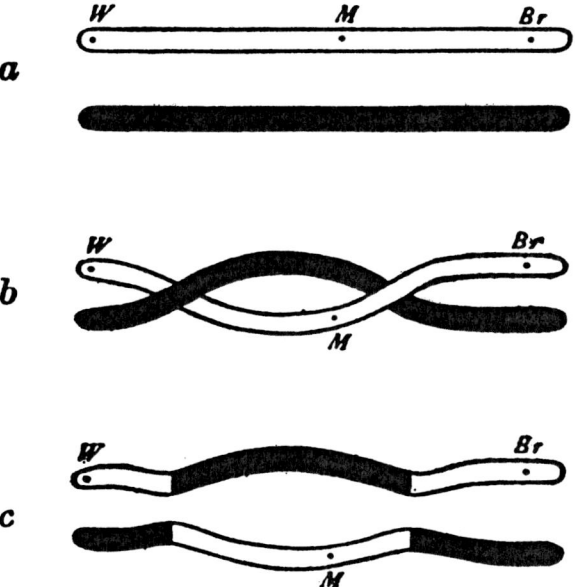

Fig. 2. Diagram to illustrate double crossing over. The white and the black rods (*a*) twist and cross at two points. Where they cross they are represented as uniting (shown in *c*). That an interchange of pieces has taken place between *W* and *Br* is demonstrated by the factor *M* having gone over to the other chromosome. The Morgan school's explanation of the "linkage" of traits as a result of the increasing probability of crossing over between factors at increasing distances along the chromosome. On the simplest model (ignoring interference), this model also predicts linkage relations like that of Trow's Law. From Morgan, Sturtevant, Muller, and Bridges, 1915, fig. 25, p. 62. Reproduced through courtesy of Henry Holt and Company.

had to contend with. It would have invited confusion for the Drosophila group to use a formula associated with the reduplication theory — even heuristically, to serve a dydactic purpose — as they could have in their arguments with Castle. Thus it may not have been open to them, as it is today to us, to use it in explaining their theory.

To understand the rationale for "Trow's formula" from the point of view of the linear linkage model, suppose that factors A, B, and C (or *W*, *M*, and *Br*, in Figure 2) are arranged in that order in a linear chromosome, that breaks and recombinations occur randomly along its length, and that the occurrence of any one event is probabilistically independent of any other. Suppose that m is the probability of a break between A and B, and n is the probability of a break between B and C. If the two breakage events are independent, then (simply from the multiplicative rule for the calculation of the probability of compound events from their individual probabilities) the number of cases of a double break between A and C should be mn. These mn double breaks will involve $2mn$ single breaks. (See formula 1, Muller 1916—II, pp. 286—6). But double breaks between A and C will be scored as no breaks, since A and C will end up on the same side. (See Figure 2, from Morgan et al., 1915, p. 62, and formula 2, Muller 1916—II, pp. 286—7.) Thus the number of observed breaks between A and C — including all of the single breaks minus the expected number which are involved in double breaks — should be:[4]

$$f(AC) = m + n - 2mn. \tag{1}$$

As both Sturtevant (1914) and Haldane (1919) show, this is algebraically equivalent to Trow's formula — something more readily seen from Trow's diagram (Figure 1) than from Trow's own discussion. I think it is fair to say that all hands (in this company) accept Trow's formula as giving the expected observed recombination frequency in the absence of interference. Indeed, it is one of the key criticisms of the Morgan group of the reduplication theory (see, e.g., Bridges, 1915) that it cannot deal with interference. And that is where we will leave it for the time being.

IV. INTERFERENCE AND COINCIDENCE: A
MECHANISM-AND-DATA-DRIVEN DEVELOPMENT OF THEORY

Muller's dissertation, published as 'The Mechanism of Crossing Over' in four parts in The American Naturalist in 1916, was a detailed study of the mechanisms of recombination and interference. It was this (or these) experiment(s) that provided the data used so effectively by the Drosophila group against Castle (see Muller, 1920, and Wimsatt, 1987). Its experimental center-piece was the combination of 12 mutant factors from the X-chromosome in a single strain so as to be able to detect with high accuracy (rather than to infer) all of the single and multiple crossovers that occur, and thereby to be able to measure linkage distances and interference effects free of errors due to undetected crossovers and confounding effects derived from combining results from different experiments. *To my knowledge, no larger number of (morphological) mutant loci have been followed in a single experiment, before or since!*[5] This was no mean task, because combining a number of mutants — each detrimental — in the same organisms meant enormous net decreases in viability and fecundity, and it was difficult (and took some genetic planning) even to construct and maintain the stocks for the experiment. (One wonders whether the difficulties in maintaining these stocks may have in part suggested Muller's later influential views on genetic load — see Muller, 1950, and also Crow, 1987). These decreases could both make it impossible to collect all of the necessary classes of flies, and also, by affecting the ratios collected, systematically bias the estimates of recombination frequency. This required sometimes quite convoluted experimental designs to avoid working with multiple mutant homezygotes.

It also led Muller to derive formulas which eliminated the effects of one of the main confounding variables in such experiments. From Mendel on, hybridists recognized that different fitnesses of the different types (if expressed before the ratios of types were measured) could make it difficult to accurately assess the ratios of types produced by segregation in order to figure out the genetic basis of the different genotypes. (Mendel recognized this in his statement at the beginning of his section on the selection of experimental plants that ". . . there should be no marked disturbances in the fertility of the hybrids and their offspring in successive generations . . . Occasional forms with reduced fertility or complete sterility, which occur among the offspring of many hybrids, would render the experiments very difficult or defeat them entirely." (Mendel, 1866 in Stern and Sherwood, 1967, p. 3). With linkage experiments, since the expected ratios are no longer confined to ratios of small integers, the possible expectations are less constrained, and the problem of detecting and eliminating viability effects becomes even more acute.

Muller's formulas allowed cancellation of differential viability effects in the estimation of recombination fractions between any two factors, **A** and **B**. His method involved a slight modification of the standard one. He begins by making two contrary crosses (homozygous types **AB** x **ab,** and **Ab** x **aB**) and backcrossing the F_1 and observing the ratios of the four types in these two experiments. (With no viability effects, either of the contrary crosses would have done equally well.) He gives 4 different formulas, each using 4 of these 8 different quantities, with the recommendation that the formula should be used ". . . which contains the largest number of individuals in its smallest class, for this would usually have the least probable error" (1916—III, p. 353). His method made no assumptions about the independence (or multiplicative character) of the effect(s) of the different mutants, and required only the assumption that the same phenotype had the same viability in the two experiments. (The idea of using contrary crosses to increase reliability is commonly credited to R. A. Fisher, who re-invented it some time later. Fisher, 1949, p. 215.)[6] Muller was no mathematician, but he had a working facility with algebra, an inventive conceptual insight, and a sense of experimental design which many more quantitative scientists would have been proud to own. This kind of invention of statistical procedures which suggested what experiments should be done and what data from them should be used — and how — is theory too, but not the same kind of theory as Haldane's model building. It is theory from the trenches, or theory constructed by statisticians for use in the trenches.

Muller's ultimate aim with these experiments was to determine accurately the magnitude of the effects of interference, and thereby, with what other correlative cytological information and mechanical hypotheses he could muster, decide between alternative causal hypotheses as to the mechanisms of recombination and interference. As it turned out, even his experimental results were not good enough to decide between the various hypotheses — and the nature of these hypotheses changed significantly after the development of the four-strand theory of crossing over in any case. (See, e.g., Weinstein, 1935, and Owen, 1950.) But the most lasting result of Muller's work was none of the above.

Along the way, Muller developed what is sometimes referred to as the theory of coincidence, but which is better described as a measure of interference and a set of procedures for calculating it. This approach saw occasional elaboration, but was at least passed down and treated in virtually all relevant intellectual lineages from the Drosophila group.[7] Together with the mechanical hypotheses which he favored (e.g., of a modal distance between crossovers produced by chromosome rigidity), it constituted the analogue for the Drosophila group of Haldane's causal models and equations. But where Haldane provided differential equations and theoretically derived Poisson distributions, Muller provided statistical procedures for apportioning and summing probabilities of crossing over from real data on crossing over, and coincidence, which was a measure of interference — the ratio between observed and expected crossovers as a function of map distance from the last crossover. The difference between their approaches was the difference between experimentalists who wondered how to give appropriate theoretical treatment to their data, and a theoretician, who had derived the appropriate form of an equation from an *a priori* idealized causal model, and looked to data to find appropriate parameter values for his model. It is the contrast between "bottom up" versus "top down" theorizing at its best. Perhaps not surprisingly to anyone who has ever looked at cases like this in detail (but quite at variance with what I was led to expect in my training as a philosopher) the result is not convergence or even collision, as much as two ships passing in the night.

V. A CONFUSING PLETHORA OF POSSIBLE MECHANISMS

The phenomenon of interference is first notèd in print (though not by that term) by Sturtevant in 1913. The term was apparently first used by Muller, and is in wide use among the Drosophila group by 1915 (see *MMH*, p. 64.) Bridges (1915, p. 18) says of the process that it was "... originally deduced by Muller and Sturtevant from a consideration of linkage as a chromosome process", and goes on to say that it is "... a corollary of the chromosome hypothesis, [but] almost unexplainable on any other view of linkage." Bridges also there makes the first use I have been able to find of the term 'coincidence': "In order to obtain a convenient index, the converse of interference, namely coincidence, is calculated as the percentage which the observed double crossover class is of the expected value." (1915, pp. 18—19.)

Interference is the opening topic of the second paper in Muller's four part series of 1916, and coincidence is soon (p. 288) introduced as the appropriate measure of interference. Muller is careful to point out in defining it, that coincidence should really be called *relative* coincidence because it is the ratio of double (co-*incident*) crossovers that actually occur to that number expected if there were no interference. Complete interference (no double crossovers) gets a coincidence of 0, no interference gets 1, and if double crossovers were to turn out to be *more* frequent than expected (for a coincidence greater than 1), "this would be 'negative interference', for as coincidence increases, interference decreases." (p. 289.)[8]

As noted above, this has a mildly ironic feature which is common with scientific idealizations where the "true state of nature" is defined in terms of deviations from an incorrect idealization (see Wimsatt, 1987). In this case, coincidence is defined in terms of the mathematical expectation of double crossovers, or "Trow's formula" which they have gone to some pains to disavow. Thus, on p. 285, Muller expresses the view that ". . . this latter [i.e., Trow's] scheme would not be that expected on the method of crossing over proposed by Janssens and followed by Morgan . . ." because it would permit crossovers to occur arbitrarily close together. He elaborates this further on p. 289:

On Janssens' theory that crossing-over takes place in the strepsinema stage, when the chromosomes are twisted in loose loops, crossing-over would very seldom take place at two points very near together, for this would require a tight twisting of the chromosomes. Accordingly, on this theory interference was to be expected; furthermore it would be expected that interference was very great between crossings-over that were in neighboring regions; but between crossings-over that were further apart there should be little or no interference. The results were according to this expectation . . .

We will see below that this is exactly the behavior expected on Haldane's most sophisticated causal model — that of equation (5). But Muller does not provide an analytic function from map distance to recombination frequency which would specify how (i.e., over what map distances) interference should fall off from its maximum value (signifying complete interference) to nothing. This is the starting point, rather than the terminus of Muller's discussion. Ultimately, Muller's empirically derived coincidence curves are a much more detailed picture of the behavior of interference with distance than Haldane provides. And Muller, unlike Haldane, appears to recognize at least the possibility of negative interference — a possibility which springs from his detailed consideration of mechanisms, which Haldane considers in his 1919 paper, but very superficially by comparison. The deepest difference, perhaps, is that Muller is committed to dealing with the mechanisms of recombination in all of their complexity *first*, and then letting the formalisms fall where they may.

There follows an extended discussion of when in the meiotic cycle crossing over could plausibly occur, including the implications for the mechanism of crossing over that recombination does not always appear to occur with intertwining of the threads; whether breakage accompanies loops, whether the loops be small (tight) or large, whether they all must be of the same length, be of a modal length with variation about that mode, or be unconstrained with respect to length; considering the effects of the changes in length and thickness of the threads at different points in the cell cycle; whether fusion precedes breakage of the chromosomes or conversely; and whether the twisting is apparent or real. In this, Muller artfully combines discussion of the mechanical plausibility of various mechanisms with their consistency with genetic and cytological observations. He considers a dizzying array of possibilities. In the course of this tour, he advances the (basically modern) hypothesis that recombination takes place during or soon after the amphitene stage (when the chromosomes are uncoiled as thin threads) rather than at the (thicker, coiled) strepsinema stage as Janssens had supposed — a hypothesis which he finds to be required by the observed *genetic* precision of recombination. But this still does not decide between different models of interference.

While discussing the "tight loop" alternative, Muller proposes a structuring of the problem which is of greater generality — and which emerges with some difficulty, as indicated by the parenthetic interruptions:

> . . . it will be helpful to bear in mind that crossing over can be divided into just three essential processes — a bending of the chromosomes around one another, a breaking of the threads, and then a fusion of adjoining pieces (or, perhaps, the fusion of the homologous chromosomes comes first, and then the breaking of the chromosomes at that point.) It follows from this that interference must in any case be due to one of the following three general causes: (1) Either the chromosomes are not likely to bend across each other twice at points near together (i.e., the loops tend to be long), or (2) breakage at one point for some reason interferes with another breakage nearby (even though the threads are crossed at both of those points), or (3) a fusion of chromosomes at one point in some way interferes with fusion of threads which are crossed in a neighboring region. (Muller, 1916, pp. 293—294.)

He later rejects the third alternative, and focusses on the first two. These become "schemes I and II" in subsequent discussion — and, as near as I can tell, also basically become loosely specified as "loops which are long because of chromosome rigidity", and "tight loops with breaks locally reducing tension and preventing other breaks nearby". He then sets forth to devise tests to decide among these three (but primarily among the first two) possibilities, *and for this reason* (p. 295) undertakes the complex experiments involving many loci studied simultaneously and devises his account of coincidence. *It is crucial in what follows to keep this in mind* — for this account is an instrumental set of procedures for processing data designed to answer the question as to which of several competing mechanical explanations was correct — it was not *itself* that explanation. By contrast, Haldane's theory was an explanatory idealization which utilized mechanical processes to derive equations which would predict and explain recombination frequency as a function of the mechanical properties of the chromosome and the map distance between the factors involved. It either was — or was extremely closely linked to — that explanation. *For all of this, Muller was no more of an instrumentalist than Haldane was.* In coincidence, he had an instrument, and a very good one, for the task at hand.

In looking at the first two schemes for the mechanical explanation of recombination, Muller sees a significant possible difference: If the loop length is long, and there is a modal distance (even with variation around it if the variation is not too great), then coincidence should follow a definite pattern with increasing distance:

... For small distances, the relative coincidence would be very small (interference high), for longer distances, much greater, and with still longer distances coincidence would fall again (interference would rise). For distances double or triple the length of the loop — if the chromosome were as long as that — coincidence would rise once more. Secondly, on the "loop explanation" just outlined, coincidence should, at the modal distance, rise above the 100 percent level, for crossing-over would occur at a given point (K) more often in those cases when there is crossing-over at another point (I) lying at the modal distance from K than in the average case. Of course it might be, however, that there was no modal length of loop — that although short lengths of loop were infrequent, all loops above a given size were equally frequent, or that the longer the loop, the more frequent it tended to be. In the former case coincidence would rise to a certain level, as distance between the two points of crossing-over considered increased, and would after that remain constant; in the latter case, it would rise progressively, and might or might not reach or pass the 100 percent level. (Muller, 1916, pp. 295—6.)

Muller's prose is highly reticulated at this point (indeed this is one of the crisper formulations in this section!). This fits the complexity of the situation when there are few facts he can pin down or few alternatives he can definitively rule out, so that even when he starts to talk about the consequences of having a modal distance between crossovers, his statements are immediately followed by qualifications exploding in a branching array of alternatives. Haldane (1919b, p. 300) actually embraces something like one of these alternatives — with a lower bound on loop size and loops above that size equally frequent — in his discussion of equation (4) below, but with no apparent rationale other

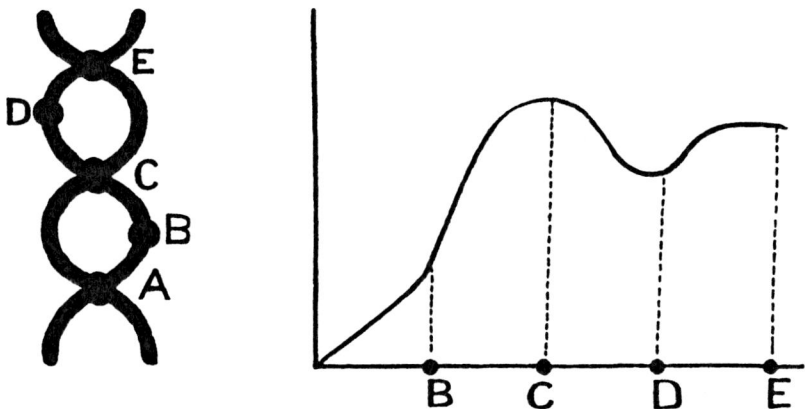

Diagram to show possible coincidence relations on schemes I and II. The chromosomes are represented as crossing-over at A, and twisting in loops of the most usual (modal) length. It will be seen that a crossing-over at A will rarely coincide with one nearby — at B — since then the chrosomomes would have to twist in loops much smaller than the modal length. But it will often coincide with one at C, seldom with one at D, and often again with one at E. The *relative coincidence* of crossing-over at various points on this chromosome with crossing-over at A is shown in the curve on the right.

Fig. 3. Muller's predictions of a "modal distance" for highest frequency of crossing over. From Muller, 1916, his fig. 8, p. 298. Reproduced through courtesy of *The American Naturalist* and the University of Chicago Press.

than that it could produce behavior approximating that equation. The picture expected for long loops with a modal distance between cross-overs is diagrammed in Muller's first presentation of a coincidence curve on p. 298 (Figure 3).

The important thing to note, however, is that he now seems to have some detailed qualitative predictions about the behavior of the relative coincidence measure as a function of distance, on different hypotheses (with and without modal distances in different favors), for long loops. He goes on to argue that on the "tight loop" hypothesis (which he now confusingly calls scheme III — perhaps because its behavior corresponds to the third alternative in the quote) you would expect with increasing distance to find monotonic increase of coincidence towards some asymptotic value less than or equal to 100%. This is again almost immediately qualified (footnote 2, p. 296) with a suggestion that it could go slightly above 100% if the inhibition of nearby crossovers made the occurrence of more distant crossovers more likely, but that qualification is further qualified with a comment on how *those* cases could be distinguished from others with which they might (readily!) be confused! This wandering tour is in turn followed (p. 297) by a discussion of the mysterious recombination modifiers he has found in chromosome III, and Sturtevant in chromosome II (later revealed — after 13 years of study — as inversions, see Sturtevant 1919, 1921, 1926, 1931), and the hope is expressed that they may themselves shed, or can be used as tools to shed, some light on the process. Similar discussions, proposals, and hopes follow for essentially all other then known possible sources of variation in linkage.

The net effect of all of this is to make it seem quite worthwhile to do the experiments necessary to construct the coincidence curves to distinguish among the various causal hypotheses. One is also convinced that, unless the results are so crisp or extreme as to be unequivocal, there might well be several alternative hypotheses still remaining (these reached, of course, on different paths from different starting points, several qualifications deep). In other words, one gets enormous respect for the complexity of the task of finding out what is going on, and for the potential complexity of the process — something which, as we will see, does not look nearly as intimidating from the perspective of Haldane's paper. There is also the idea (which I have not emphasized, but comes through in Muller's discussion) that the mechanisms might differ, at least somewhat, with different kinds of organisms, or under different conditions, and that in any case there is enormous potential for such variation because of the complexity of the process.

There is another curious difference however in Muller's presentation of coincidence that also muddies the water in this neat equation, and this is worth attending to, for it nearly illustrates the differences between their "bottom up" or "data driven" mode of theory construction and that of the "top down" approach of Haldane. All of Haldane's curves are constructed as continuous functions from the two cases — complete interference, and no interference — that he could solve for analytically. The empirical data are graphed against these curves, and a "best fit" curve is derived with x expressed as a function of y as the linear weighting of these two functions, each converted to the $x = f(y)$ form.[9] Because Muller's method of constructing coincidence curves generated a curve that was derived directly from the data, his methods were discrete rather than continuous, and reflected what he could do with the data that he had.

VI. MULLER'S CONSTRUCTION OF COINCIDENCE CURVES

The determination of interference or coincidence curves raised a new problem that construction of linkage maps did not raise directly. In linkage maps, factors were located in terms of relative numbers of recombination events occurring between them. For this it was not necessary to know exactly *where* the crossovers had occurred. But *to estimate how coincidence varied with distance, the distance between crossovers was required.* And of course, a crossover could not be located more accurately than bounding it between the closest mutant on each side. The distance between two crossovers, each located imprecisely in this way, could vary by up to the total width of the two intervals. Tracking mutants at a number of loci simultaneously as Muller did gives a number of signposts to act as bounding markers. But even for his best case — the X chromosome — 12 mutants randomly located in a linkage map some 65 units long leaves a lot of slop, in some cases (the spacing between factor *cl* and *v*) as large as 16 map units![10] One could not expect very crisp results from this procedure. It must have been like trying to thread a needle with a pair of blacksmith's tongs!

Since one couldn't know exactly where the crossovers had occurred. Muller proceeded to average their locations over the possible ones in a way that made as few assumptions as possible. Thus, for example, given that one crossover had occurred between *cl* and *v*, and another between *s* and *r* (at map locations of 13, 29; 37, 49; to the nearest whole units respectively), the crossovers could not have been closer together than 8 units — the distance between 29 and 37, and not further apart than 36 units — the distance between 13 and 49. (He picks the worst case as his example by picking crossovers in the two largest intervals, which are also quite close together, giving an enormous percentage error! See his map of the X chromosome on p. 421.) His procedure is explained via analysis of this example, and also diagrammed in Figures 4 through 7.[11]

Muller's analysis proceeded as follows:

(1) Of the 712 flies produced from parents carrying the requisite chromosomes, a total of 30 had double cross-overs. (See his table on p.

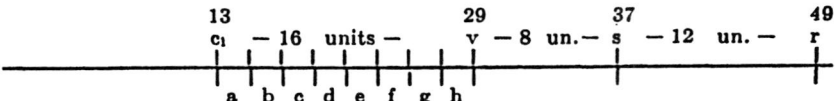

Fig. 4. Diagram to indicate relevant distances between factors and size of units Muller used to illustrate his calculation of the probability distribution for a second cross-over as a function of distance, for a case in which one cross-over had occurred between c_1 and v and a second between s and r. From Muller, 1916, unlabelled diagram, p. 426. Reproduced through courtesy of *The American Naturalist* and the University of Chicago Press.

365). Each fly that had a "double" thus contributed $1/712 = 0.0014$ to the frequency of "doubles" in the total population of flies.

(2) The doubles are classified by the intervals in which they occur. (There were no cases of doubles observed in the same interval — and could not have been since he was using all markers available to get the 11 intervals, and could not have detected an intra-interval double.) With 12 mutants, there are 11 intervals, and thus 55 possible pairs of intervals. The 30 doubles that did occur fell into 14 of these classes, with from 1 to 8 doubles per class. For each class, a probability distribution for the distance between the two crossovers had to be derived (as described in (3)—(5) below), and this distribution was then multiplied by the number of doubles in that class and the contribution per double to the total frequency (i.e., for this experiment, 0.0014).

(3) For each type of double, the uncertainty over the exact location of the two crossovers meant that their contribution had to be divided among the possible distances at which it could have occurred. To do this, it was necessary to pick a finite distance increment Δd, and locate things falling in between one distance value d and that value plus the increment $d + \Delta d$, over the range of distance intervals. As Muller notes, smaller increments would produce a smoother curve, but would require more calculation. For his example (see Figure 4), Muller picks $\Delta d = 2$,[12] for which crossovers between cl and v, and between s and r, could have occurred in one of 14 intervals, as close together as 8 units or as far part as 36 units. There were also (coincidentally) 8 double crossovers of this type — i.e., involving these two particular chromosome segments.

(4) But not all distances were equally likely. Thus, while there was only one way in which the two crossovers could have been 8 units apart

(by sitting right up against v and s, the near boundaries of the two intervals) and similarly only one way in which they could have been 36 units apart (by being plastered up against cl and r, the outer boundaries of their respective intervals), for any values in between those there was more than one way of fitting a distance of that length so that both of its ends fell in their respective intervals. Thus distances of 10 (i.e., 10—12) had two ways they could fit in, because they had an extra length of 2 units which they could shift left or right. Similarly, cutting the distance to 32—34 allowed a shift between two positions without pushing either end outside of its interval. Distances of 12—14 or 30—32 had three possible ways, etc.

(5) Thus the shape of the distribution of probabilities is a symmetric bar graph with a series of 14 bars, each two units wide, between $d = 8$ and $d = 36$. This distribution climbs linearly between 8 and 20 (in sawtooth fashion because of the finite intervals, though his graphs suggest that Muller smoothed them as in Figure 5, below), is flat between 20 and 24, and then declines linearly between 24 and 36. (It

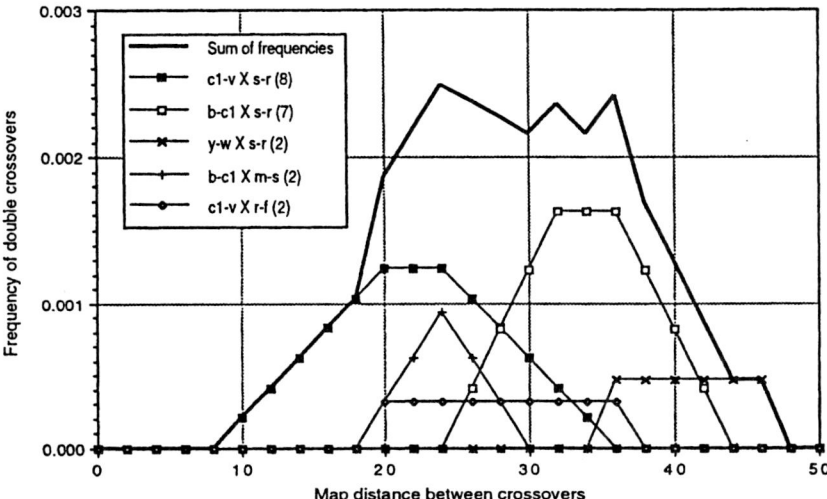

Fig. 5. Muller's construction of frequency curves for double crossovers of uncertain location, including the individual curves and their sum for the 6 highest frequency classes of data. Completion of this for all classes should give the empirical curve found in Figure 6. (Reconstruction by the author.)

has a flat top unless both intervals are of the same width, because in this region, each possible increment or decrement in distance which keeps a "leg" in the more tightly bounded region loses a possibility at one end of the less tightly bounded region for each one it gains at the other end.) It is $[(20 - 8)/2] = 6$ units high, and the area of a unit is such that the total area of the bar graph is 0.0014 times the number of doubles in that class (i.e., 8), or 0.0112. (This distribution is graphed (using larger black squares), together with the distributions for the four other classes of highest frequency and their cumulative sum in Figure 6. The number of cases for each class is given (in parentheses) after each pair of intervals defining that class.)

(6) This process is repeated for each of the 14 double crossover classes, in each case dividing its contribution among each of the possible separations and pairs of locations consistent with the bounding intervals, and producing a curve for each (or more likely an array of numbers for the different distance increments).

(7) The final curve computed from the data is the sum of these curves (or sum of the numbers for each distance increment). A partial sum for the 5 most frequent classes is given as illustration of the method above in Figure 6, and Muller's (1916, p. 424) resultant curve (utilizing all 14 classes) for the observed frequency of double crossing over in the first chromosome is given by the solid line in Figure 6.[13]

In this composite curve, the sharper peaks are produced by pairs of individual crossovers both of whose locations are known relatively accurately due to short intervals between the bounding mutations, whereas the smoothing is produced by uncertainty as to location. (Contrast the last two curves, using the '+' and 'diamond' symbols, in Figure 5. For the '+' curve, the intervals $b - cl$ and $m - s$ are of the same, relatively narrow width (rounded to 6 units) producing a distribution with a width of 12 at the base and a relatively high point-peak, while the 'diamond' curve is produced by intervals of $cl - v$ and $r - f$, of (rounded) lengths 16 and 2, respectively, produce the lower flat plateau of base 20 and top 16 (both 2 too long from round-off error). Relatively speaking, the example discussed at length here involves more smoothing than any other curve, but with 8 doubles (compared to the 2 doubles of the '+' and 'diamond' curves), the higher amplitude and area of this curve hides its smoothing effect.)

As Muller himself notes (p. 427), he used a rather (indeed maximally!) conservative averaging procedure to produce the curve — his procedure assumes nothing about interference which is not directly

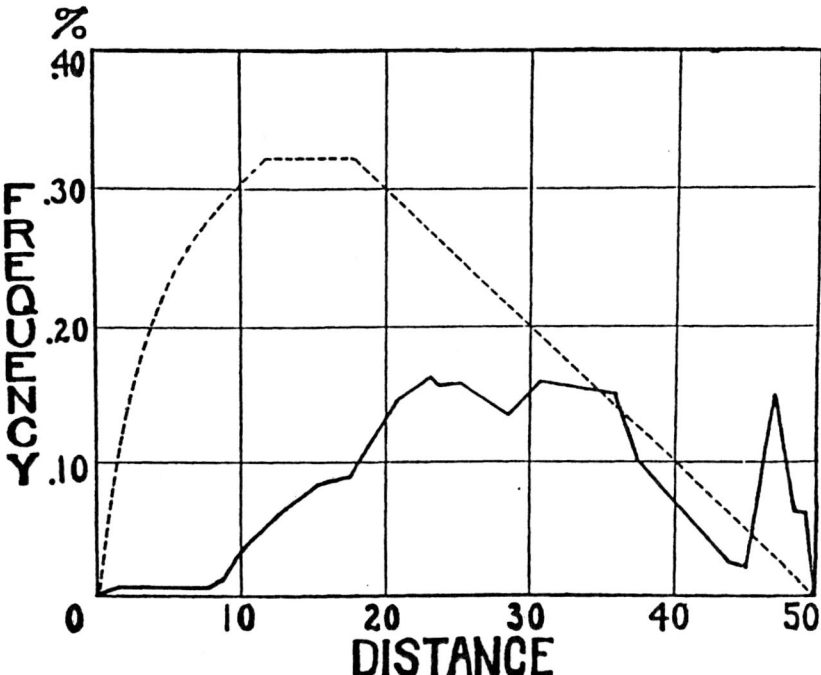

Fig. 6. Ideal and "observed" (constructed by the method of Figure 5) curves for frequency of double crossovers as a function of separation distance for the X chromosome. From Muller, 1916, his fig. 12, p. 424. Reproduced through courtesy of *The American Naturalist* and the University of Chicago Press.

entailed by the data. His procedure (may) show interference in terms of how the crossovers are distributed *among* the various regions, but he cannot show (and does not assume, in his analysis) the action of interference in the placement of crossovers *within* regions.[14] If the minimum distance between crossovers were actually 14 (Weinstein, 1918, p. 54, says that in the X-chromosome, crossovers have never been observed closer than 13.5 units apart, and this again would have to be a lower bound), Muller's procedure would, in the above example, assign an increasing probability of double crossovers starting at 8, *simply because he can only rule out as impossible distances which are still shorter*.[15] If the probabilities of crossing over are not in fact uniform in the way required by his construction (as they surely are not), he would tend to underestimate the effects of interference (i.e., over-estimate coincidence) at shorter distances.

As Muller does not note, *the same bias would apply symmetrically at the other end.* Again, in the above example, a crossover at 14 would have its probability weight distributed out to 36 (or 42 in "standard units" — see preceding footnote), because one could rule out as impossible only distances which were still longer. Thus, if there is a modal distance and consequent increase in coincidence (even to the point where it could show "negative interference") near integral multiples of that distance — his averaging procedure would tend to smooth out the curve so as to underestimate coincidence in that region. This is especially important, because the evidence was more equivocal on this point than it was for the existence of interference at short distances: *Muller's procedure would improperly weaken the evidence for a maximum of coincidence near the modal distance and its integral multiples.*[16]

The net effect of Muller's procedure then is to be conservative with respect to confirmation of his favored hypothesis, because it both overestimates coincidence at shorter distances, and would underestimate it near modal distances. This admirable restraint may have led to more scepticism than was justified towards his theory.

In addition to the data curve giving frequency of crossover as a function of distance, Muller needs a "theoretical" curve, giving the expected probability of crossover in the absence of interference, to estimate his "relative coincidence". He derives this in the same manner used to get Trow's function, but instead of going in Haldane's direction (which would presumably be to utilize a Poisson distribution to estimate the probability of a second crossover as a function of distance), he is particularly concerned to make sure that his generation of the curve parallels as closely as possible the way in which he generated the data curve:

. . . To make comparison with the other curve legitimate, it had to be constructed by the same method — namely by making a composite of the individual curves, each of which represented the probabilities for a certain type of double cross-over — only instead of using the observed number of double cross-overs of the different types, in constructing it, it was necessary to use the number of different cross-overs of the different types if there had been no interference. (This curve thus represents the results of a chance distribution both among and within the various regions.) (Muller, 1916, p. 428.)

This curve would have been a fair amount more work, because he had to use the 11 actual interval-lengths specified by his 12 marker genes. His data involved 30 crossovers falling into 14 of the 55 classes. For the data, he thus had to evaluate and add together only 14 distributions. To generate his expected curve, for all 55 pairs of intervals of lengths m_i and n_j, he presumably calculated the product $m_i n_j$ (the expected frequency of that type), generated their distributions (each of area $m_i n_j$, respectively), and added them all together. His ideal expected curve is represented by the dotted line in Figure 6.[17]

Given the data curve and the curve for theoretical expectation, the relative coincidence curve is gotten straightforwardly by dividing the observed by the expected values for crossovers as a function of distance. This graph then can be generated directly from the two curves in Figure 6. Muller gives this in his (and our) Figure 7, p. 429.

After all of this work, Muller is surprisingly reticent about drawing much of any conclusions from it at all. He notes the gradual rise in coincidence (or decline in interference), but refuses to draw any conclusions from the peak at about 35 centimorgans on grounds that there are few crossovers at greater separations, and 'the dotted portion of the curve (from the peak on) he regards as untrustworthy. He says that in any case, if the modal distance hypothesis were true, one would expect a curve which was symmetric about the peak, and a peak value of

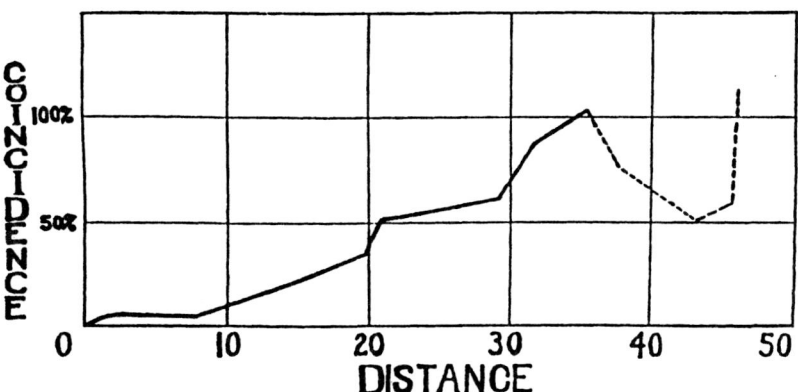

Fig. 7. Coincidence curve for the X-chromosome: constructed as the ratio of the two curves in Figure 6. From Muller, 1916, his fig. 13, p. 429. Reproduced through courtesy of *The American Naturalist* and the University of Chicago Press.

coincidence significantly above 1, and since neither condition appears to obtain, one can draw no conclusions.[18]

Muller promises in this paper to do coincidence curves for the other two major chromosomes when he has sufficient data, but I am aware of no published results by Muller on either of them. It is possible that the finicky and laboriously assembled stocks for this experiment did not survive the trip to Rice university — or that he did not take them, when he moved there in the fall of 1915, having rushed to finish his dissertation in time to take the job (but too late to receive his degree before 1916 — Carlson, 1981, p. 92). Muller's work on coincidence was further extended in an oft cited study by Alexander Weinstein in 1918, and Weinstein went on to substantial further work on models of recombination and interference. (See, e.g., Weinstein's work on the four stranded model in 1935.)

VII. HALDANE'S CAUSAL MODELS

Haldane's 1919 paper contained many surprises for me when I first looked at it. I expected to find Haldane inventing and giving a pretty derivation of the formula he is credited with, but instead, found a different and far richer trove. First of all, the paper in which he proposes linkage models and mapping functions (1919b) is immediately preceded by another general statistical paper (1919a) in which he does an error analysis of the calculation of linkage values, and proposes how it should best be done — techniques which he occasionally puts to use in the oft-cited paper. Secondly, his better known (1919b) paper contains far more than advertised — an elegant piece of theoretical model-building involving three different models from which he derives a fourth, and a separate empirical curve-fitted equation based on two of them for prediction. These various models are interspersed with applications of them to evaluate the status of two versions of the reduplication theory, hypotheses about recombination attributed to Morgan and Bridges,[19] and estimates of the total map length of the X-chromosome.

Haldane begins his discussion with presentation of Trow's formula in the form we saw above as equation (1). What we know as Haldane's mapping function is derived as a continuous function relating map distance to recombination frequency, assuming recombination events are independent, and in which an odd number of recombinations between two factors are scored as a single crossover, and an even number is scored as no crossovers. (Haldane actually gives two derivations of this result, first by differentiating equation (5) below and solving for $p = 2$, and secondly by summing terms for crossovers which are observable (i.e., odd numbered) in the Poisson distribution for the expected different numbers of crossovers.) Where x is map distance, any y is observed recombination frequency, Haldane's mapping function is given by equation (2) [his equation (3), p. 303]:

$$y = (1/2) [1 - e^{-2x}]; \quad \text{or} \quad x = -(1/2) \log_e[1 - 2y]. \quad \text{(2a, 2b)}$$

Of the two forms Haldane gives, the first is preferable as a causal equation, since in it the (causally) dependent variable (recombination frequency) is expressed as a function of the (causally) independent variable (map distance). The relevance of the second form will appear later, when we discuss Haldane's predictive equation.

The second surprise is that "Haldane's mapping function" represents but one of 5 models he discusses, and there is no sign that his contemporaries (or most modern citations of his paper either!) recognize the sophistication of his discussion. Haldane begins (p. 300) by discussing three different physical models of how the properties of the chromosome affect the likelihood of multiple crossing over.

In the first model, he supposes that the chromosomes are "infinitely flexible" —in which case, one assumes that the occurrence of one crossover would not affect the occurrence of any other crossovers — so that crossovers are probabilistically independent. This results in probabilities of single and double crossovers like that given above for Trow's formula, in equation (1).

VIII. HALDANE'S SECOND MODEL AND
THE "LINEARITY" ASSUMPTION

In the second model, Haldane supposes that the chromosomes are "straight rigid rods", in which case, at most 1 crossover can occur. If no more than one crossover can occur, then all crossovers are detected, and crossover frequencies and map distances should be strictly additive. On this hypothesis,

$$f(AC) = m + n. \quad \text{(3)}$$

He later argues (because of the straight additivity, p. 304) that this corresponds to the mapping function, $y = x$, which he labels (somewhat perversely, as we shall see) "Morgan and Bridges' formula" — often joined with claims or implications that it is incorrect (p. 300, see also pp. 299, 304). Although the Morgan school argued for straight additivity of recombination frequencies for small map distances, it was clear to them from the beginning (see Sturtevant, 1913, and virtually any paper by any of them since) that this relation broke down for larger distances. Indeed, their most frequent claim is not for linearity or additivity, but that recombination frequency is a *function of* or is *calculable from* map distance (e.g., Morgan *et al.* 1915, p. 65; Muller 1916, p. 201; 1920, p. 98; Morgan 1919, pp. 118—125). Furthermore, *all* of the equations Haldane considers behave approximately additively for short distances, as Haldane himself implicitly admits in claiming (pp. 305—6), that ". . . smaller crossover values . . . do not allow of much discrimination between the three equations." (See also p. 302.)

This does not quite end the matter. The Drosophila group had an additional argument for strict rather than approximate additivity for small distances in the mechanism of interference (for distances less than 5 to 20 map units depending upon the chromosome and location in it). But this is more properly seen as a second order correction on a first order Trow-like formula, than as a zero-th order failure to recognize any complexities at all. (Recall Bridges comment (1915, p. 17) cited above that interference is to linkage just as linkage is to free assortment, in that interference involves denial of independence of crossing over just as linkage involves denial of independence of assortment.) Unfortunately, Haldane's references suggest the simple-minded view.

Misunderstandings of the nature of the chromosomal maps (by lesser minds than Haldane, though!) were legion throughout this period, and the Drosophila group often had to deny positions misattributed to them by others when they had already clearly stated their views many times in print. It was in fact a very sophisticated conception, whose subtleties were often missed. This misattribution is doubly ironic, since not only have they been explicit and clear in denying the first view (see, e.g., pp. 21—23 in Morgan and Bridges, 1916, the very work Haldane cites), but they are at that same time (1919—20) going through an extended argument with Castle (e.g., 1919a, b), who in fact *does* hold the view they are being attacked for. If the linear mapping function was to be attributed to anyone, it should have been Castle. All of the members of the group in different combinations argued that a linear mapping — and Castle's interpretations based upon it — will not work. Plough, Sturtevant, Bridges, Morgan, Metz and Muller between them contributed 9 authorships of 5 papers criticizing Castle within little more than a year — see especially Muller (1920) or Wimsatt (1987) for further details.

IX. HALDANE' THIRD MODEL AND HIS CAUSAL META-MODEL

In the third model, Haldane supposes that double crossovers are impossible, but that each one which would have occurred on the first model is replaced by a single crossover — something which he suggests might be approximately right if the chromosomes were "semi-rigid" (p. 307) or could not produce loops of less than a given length. In this case, he suggests.

$$f(AC) = m + n - mn. \qquad (4)$$

For these models then, the linear dependency of (3) must be corrected for the possibility of multiple crossovers, which would yield equation (1) if the locations of the crossovers were statistically independent or random. But, as Sturtevant (1913) first pointed out, and Muller (1916) documented in detail, they are not independent, but interfere with one another at close range, forcing a correction of (1) in the direction of (4). But while (4) is a move in the right direction from either (1) or (3), it is the least well motivated of the three models, and seems somewhat *ad hoc*.

One might think that these three models represent alternative models of the chromosomes, which Haldane wants to test. It is interesting that in each of these cases, he does not content himself with a mathematical expression, but attempts to give a physical rationale or model for why the chromosomes should behave in that way. The mode of reasoning is rather like that of a mathematical physicist considering the consequences of whether a given physical system is like one of a variety of ideal systems (compare, e.g., Cartwright, 1983, or Wimsatt, 1987). But in fact, it looks like Haldane is being more sophisticated yet. He presents a table of the data of all known 3-point crosses (p. 301, taken from Morgan and Bridges, 1916), classified by which of these three models it appears to fit, juxtaposed with calculated values from the three models, from which it emerges that the best fit moves in sequence through the three models as the distance, $m + n$, increases. (The table seems designed in fact to illustrate this point.) These three models represent points in a continuum of models in which the chromosomes range from being infinitely flexible to completely rigid. For this he constructs a "meta-model" in which he introduces a parameter, p, which captures all three of the prior models, and suggests that all of them are right — each for different map distances:

> It has been shown above that if A, B, and C are three factors whose loci lie in that order in the same chromosome, and if m and n are the cross-over values for A, B, and B, C, respectively, then the value for A and C is $m + n - pmn$, where p is a number between 0 and 2, increasing on the whole with m and n, and having the value 1 when $m + n =$ about 0.5. (p. 303.)

We now have a meta-model, in which p is itself a scale-dependent variable; so that $p = g(x)$:

$$f(AC) = m + n - pmn. \tag{5}$$

This suggests a very sophisticated view of the problem — one in which, for close distances, the chromosome is treated as (nearly perfectly) rigid, for very great distances as' (nearly infinitely) flexible, and for in-between distances as of intermediate rigidity — in which, in effect, the properties of a system are dependent on scale. It is what one might expect from the author of later essays on allometry — the influence of size scale on strength and other adaptive capabilities. (Thus, see his essay, "On being the right size", in Haldane , 1927.)

Of course, one could note that (4) is simply mathematically intermediate between (1) and (3) and suggest that the passage from one to the other is simply a mathematical operation with no causal or physical significance. Thus, in his discussion (p. 304) of the relation between equations (3), (1), and (4), he uses a kind of functional analogue to the Mean Value Theorem, and earlier (p. 300), he says "Hence the cross-over values for A and C should be approximately $m + n$ when m and n are small, $m + n - 2mn$ when their sum is large, and $m + n - mn$ for intermediate values." But these remarks are also consistent with a thoroughly realistic interpretation, using the mathematics to describe the changing dependencies for the causes in variation of p. Elsewhere, he gives clearly physical interpretations to the equation:

> The supporters of the reduplication theory must therefore explain the deficiency of the double cross-over classes of gametes ... On the chromosome theory this is due to the rigidity of the chromosomes, and until an equally plausible explanation on the reduplication theory is given, the chromosome theory must be considered the more probable of the two as far as the class of evidence dealt with in this paper is concerned. (p. 303.)

and, after expressing the map distance as a sum of a series of terms, c_0, c_1, c_2, c_3, . . . expressing the probabilities of n crossovers for $n = 0$, 1, 2, 3, . . . for the "infinitely flexible" case:

In practice, however, owing to the rigidity of the chromosome, the value of c_1 thus calculated is too small, and those of c_0, c_2, etc. too large. They are however more accurate for great lengths, where the rigidity of the chromosomes affects the results to a less extent. (p. 305.)

I believe that Haldane is here using a continuum of idealizations to describe causal behavior ascribed to a real (but still idealized) system. (It is real because the parameter p — in effect a rigidity-scale parameter — is left unanalyzed theoretically, but treated as an empirical parameter, which is both observed to vary between 0 and 2, and one could argue theoretically should vary between 0 and 2. It is idealized because the form of the equation is suggested by equations (1) and (3), which form two ends of a theoretically constructed and idealized continuum of systems of different rigidity, though one could not say actually what the theoretical form or causal content of the equation was without specifying the analytical form or causal dependency of p on map distance, x. See also Wimsatt, 1987.) This claim (that Haldane here intends a causally realistic but idealized model) is in part supported by contrast with what he does next, in providing an equation for predictive purposes. We will return to this after a detour in the next section to explain the relation between Muller's notion of coincidence and Haldane's general causal model.

X. THE RELATION BETWEEN
INTERFERENCE AND COINCIDENCE REVISITED

There is indeed a very direct relation, though no-one attempts to point this out. With map distances of m and n between factor pairs A, B and A, C, the expected number of double crossovers is mn. Haldane's most general equation (5) multiplies this by a factor, p, which varies with distance, starting at 0 and increasing to 2 — which it must be when there is no interference to correct for the 2 crossovers missed each time we fail to score a double. Weinstein's implicit definition of coincidence (1918, p. 153) makes the connection apparent:

If a and b are respectively the lengths of (proportions of crossing over within) the regions under consideration, the amount of double crossing-over involving both regions simultaneously is *abx*, where *x* is the coincidence.

In other words, coincidence is simply equal to $p/2$, and Muller's construction of coincidence curves is simply an empirical determination of the function $p(x)$ which Haldane leaves largely unspecified, except at its endpoints! So given this connection, how could they *fail* to communicate? (Indeed, how could Haldane have failed to cite Muller — which he does a year later (Haldane, 1920) for Muller's discussion of recombination modifiers!) Remember, however, that one must agree on the endpoints as well, and Haldane thinks that $p(x)$ approaches 2 asymptotically and monotonically from below with increasing map distances, while Muller (by his own expression of the hypothesis he favors), would expect that $p(x)$ would reach a maximum value significantly greater than 2 at the modal distance, and would subsequently oscillate (he diagrams a decaying oscillation) with a period equal to the modal distance. This is a qualitative prediction, of course, because we haven't specified the modal distance, but it already distinguishes the two hypotheses, because no metrical changes can transform monotonicity into oscillation. Furthermore, it by no means follows that the modal distance (or the function $p(x)$, either) is the same for different chromosomes, or even for different points in the same chromosome. Indeed, it follows from the"correction map" presented below (see Bridges and Morgan, 1923, and Morgan, Sturtevant, and Bridges, 1925) that they are not.

XI. HALDANE'S PREDICTIVE EQUATION

After the discussion of his causal model, Haldane shifts gears noticeably to provide a more accurate equation for predictive purposes. Noticeable in this new context, by contrast, is the absence — indeed, virtually a denial — of any physical motivation for this new equation:

To obtain a more accurate relation between x and y we may plot the curves representing equations (3) and (2) and then obtain empirically a curve lying between the two which fits the observed results as closely as possible. This has been done in the figure, where line (a) represents equation (3), line (b) represents equation (2), and curve (c) [represents equation (6),]. . . . [Equation (6)]is merely chosen to give as good a fit as possible, and has probably no theoretical significance. (p. 305.)

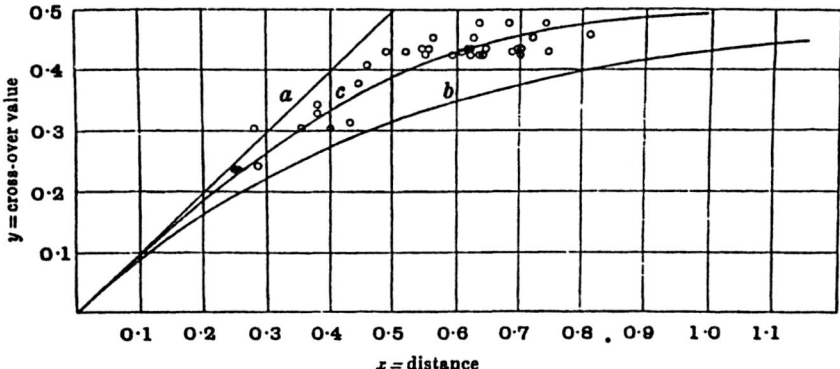

Fig. 8. Haldane's graph of the relation between (observed) frequency of crossing over and (calculated) map distance, using the data of Morgan and Bridges, 1916. Curves (a) and (b) represent equations (3) and (2) in the text, with the intermediate curve being his "best fit" predictive equation — equation (6) in the text. From Haldane, 1919, p. 307. Reproduced through courtesy of *Journal of Heredity*.

Haldane thus presents his empirical or predictive curve as inter-mediate between the two extremes represented by equations (3) and (2) — the extremes of a perfectly rigid and an infinitely flexible chromosome. It joins two other equations — equations (4) and (5) in that status. But although *all* of *them* are given a causal gloss, it is not. Indeed, a little later (p. 306), Haldane even acts as if his predictive equation has no preferential connection with the chromosome theory:

Hence the curve (c) may be taken as a fairly accurate guide to the combination of linkage values, and this remains equally true whether the chromosome theory is adopted or not.

This last aside probably would not have won him any friends in the Drosophila group, which regarded linkage mapping, justifiably, as the most powerful support they had for the chromosome theory. Haldane's equation (6) is curious in another way, because it expresses x as a function of y rather than y as a function of x, as was done for almost all of the previous equations:

$$x = 0.7y - (0.3/2)\ln(1 - 2y).\qquad(6)$$

His justification for calling this a "best fit" is twofold — first, that of the 36 data points he has (from Morgan and Bridges, 1916), 18 fall above the curve and 18 below. Indeed, the major way that he scores other models is to evaluate how they divide the points, with a highly skewed division leading to their rejection. But equation (6) is in another way a curve lying between (2) and (3): it not only has them as upper and lower bounds, but (though Haldane does not explicitly say so) it can be seen to be a linear combination of them when they are expressed with x as a function of y. Using $x = y$ for (3), and the (2b) form of (2), we have:

$$\text{equation (6)} \equiv 0.7 \text{ [equation 3]} + 0.3 \text{ [equation 2b]} \qquad (7)$$

The weights 0.7 and 0.3 were presumably chosen empirically to divide the points equally — he offers no direct justification for these values, and as he says, this equation "probably has no theoretical significance (p. 305)." Indeed, this inversion, in which x is expressed as a function of y, would seem to support the intuition that the equation is of no theoretical significance, since, although still a predictive equation, it is no longer a causal equation: saying that the map distance is a causal (not an inferential) consequence of the recombination frequency seems as meaningful as saying that the height of a flagpole is the effect of the length of its shadow.

But why should one invert the equation in this way? Why not do the linear combination of equations (2) and (3) with y expressed as a function of x? The problem is that it can't be done because the results aren't equivalent. Doing the linear combination of the equations with y expressed as a function of x would contain from equation (3) a term of form cx, where c is some positive constant between 0 and 1, and, for any value of c, would thus result in values for recombination frequency exceeding 0.5 for sufficiently long map distances, x. By contrast, any linear combination of the two curves done with x as a function of y will asymptotically approach 0.5 for recombination frequency as map length increases, which Haldane (and most subsequent theorists) obviously desired.

Haldane clearly marks this equation as intended for application by providing (p. 306, table II) a computed table of its values for use by researchers, (something he doesn't do for any of the other equations), and goes on to suggest that "... as more results accumulate, it should be possible to correct these values, which are rather uncertain for large values of x and y." (This uncertainty was not so great as to prevent him (pp. 307—8) from expressing disagreement with Morgan and Bridges' (1916) estimate of the map length of the X-chromosome, at what were surely, for him, large values of x and y!) Unfortunately, he does not always mark the significant distinction between this last predictive equation and his earlier physically motivated models — and in different contexts he will refer to "the theory" to mean sometimes one and sometimes the other.[20] Thus he refers to "theory" or "theoretical" on pp. 302 and 307 where he seems to mean the physically motivated models, and similar locutions on p. 299 and in the quote given below where he seems to mean his predictive equation:

It is believed that the above method of estimating distances will prove of considerable value when applied to comparatively long chromosomes in which factors are sparsely located, such as the second and third in Drosophila, since there is no reason to suppose that the relation arrived at between distance and cross-over value is peculiar to the sex chromosome in Drosophila. The results of investigations on these chromosomes should go far to confirm or refute the theory. (p. 308.)

Which "theory" is Haldane talking about here — his model or models or his predictive equation? I suggest that Haldane's "theory" is properly represented by equation (5) — his integration of the models expressed by equations (1), (3), and (4). But in the absence of an explicit formula relating parameter p to map distance x, Haldane cannot make predictions from this theory. He constructs equation (6) for this purpose, as a purely heuristic curve-fitting approximation to the data. (It is interesting that Kosambi in 1944, who does clearly understand the distinction between equations (5) and (6), discards (6) and takes providing an analytical formula for p in equation (5) as his task in elaborating Haldane.) Although both equations (5) and (6) are each in its own way intermediate between equations (2) and (3), they are not directly comparable to one another, and it is not clear how a confirmation or disconfirmation of (6) would bear on the truth of (5). Haldane does not make this fact clear. Given the invitation to test "the theory", anyone might well take disconfirmation of (6) as also affording a disconfirmation of the more interesting, plausible, and harder to test (5) — if they even clearly distinguished the two. I can find no unambiguous sign that anyone in the Drosophila group did, although given the focus of their attack — on the very generality of Haldane's account — they may not have needed to.

XII. THE ACCURACY OF MULLER'S DATA

Another feature of the Drosophilists' debate with Castle was also directly relevant to Haldane's analysis. This is the sensitivity of his conclusions to the quality of the data. Muller (1920) argues in detail against Castle that if the linear linkage model is under test, very accurate data is required. Castle complains that he is using their data — just as they are. Muller points out that data adequate for one purpose may not be sufficiently good for another. In particular, data with tolerable amounts of error for constructing chromosome maps (presupposing their theory) cannot be used to *test* their theory because Castle's construction method will produce a significantly non-linear arrangement (apparently confirming his theory) with very small deviations from additivity of the sort that could have a variety of other causes.[21] (It is characteristic of the Drosophila group that this is not an empty worry or a principled claim — they always have specific other causes in mind.) For all of these reasons, Muller claims that to minimize error for such a test, all of the data must be taken from the same experiment. (Muller, 1920, p. 107, see also Sturtevant, Bridges, and Morgan, 1919, p. 169.)[22]

To construct his models, Castle used the data set from Morgan and Bridges, (1916), which combined data from all experiments to date. But all of the above authors (including Morgan and Bridges) agree that their data set will not do for Castle's purposes. This is the same data set used by Haldane — perhaps reasonably, because it was the most complete reporting of data for any chromosome then available. But Muller's criticism is well founded. So what should Castle (and Haldane!) do? Muller just happens to have such an experiment handy — namely the one(s) he reported on in 1916, and he utilizes a subset of this information. (He picks, from the 12 factors originally followed, 6 which are relatively evenly spaced along the chromosome.) The others also use his data from those experiments to undercut Castle's claims. If we assume that this data represents their best data on the X-chromosome at that time, do corresponding criticisms apply to Haldane's analysis as well?

Despite Haldane's care in some other respects (e.g., marking distance estimates where the sample size was too small to give an accurate measure), they do. As Muller makes abundantly clear in his original study (1916—III), the whole point of doing a single experiment was to get the accuracy and freedom from systematic distortion required to analyze mechanisms of crossing over and interference. They had good reasons (summarized in Morgan, 1919, chapter 12) to believe that there were a number of different factors which could cause major variations in linkage. Haldane's models differed from one another primarily in the scope they gave for interference, so it seems that he too would need the greater accuracy of Muller's 1916 data. The benefits of Muller's careful experimental control are apparent in Figure 9, where his data (for the 6 of the original 12 loci that he picked out for his argument with Castle in 1920) are graphed along with those of Morgan and Bridges and Haldane's "best fit" curve for their data (equation (6)).

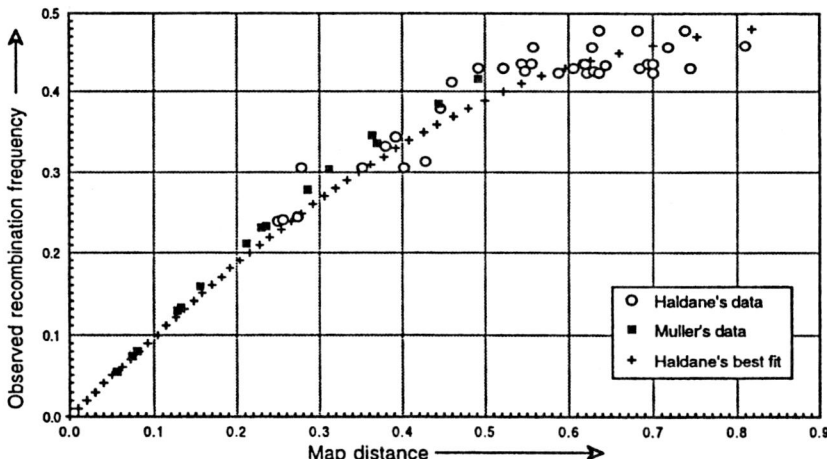

Fig. 9. Haldane's best fit curve from Figure 8, graphed together with Morgan and Bridges 1916 data used by Haldane and the data from Muller's experiments. (Graph by the author.)

While the issues are a little more complicated than indicated here, for the map distances where they overlap there is at least an order of magnitude more fluctuation (and thus perhaps two orders of magnitude more variance) in the Morgan—Bridges data used by Haldane than in Muller's data.[23] For these kinds of purposes, Muller's data seems to be enormously more accurate. Whose data would you trust!?

Even more importantly, Haldane's best fit empirical curve (equation

(6) above) deviates *systematically* from Muller's data, falling further and further below it for larger map values, so the difference is not just one of noisier data. It is important to see that this divergence is not with any of Haldane's idealized models, or with the "Haldane mapping function" which ignores interference. It is with Haldane's empirical "best fit" curve, whose form presumably *does* reflect the effects of interference. Since the curves get flatter for higher map values, this makes for very large predictive differences between Haldane's curve and any reasonable extrapolation from Muller's data.[24]

Using his equation (6), Haldane came up with an estimate of 71.5 map units for the total length of the map of the X-chromosome (p. 307), compared to Morgan and Bridges' (1916) figure of 66.2. (Haldane recognized that larger map values, with recombination frequencies closer to 0.5, would be most sensitive to changes in the mapping function, in conjunction with the possibility of correcting the values that he provides (p. 306). But this makes his disagreement with members of the Drosophila group over the length of the X-chromosome especially suspect.) Although Haldane notes the uncertainty of his estimate (p. 307) he does not relate it to this issue. I suspect that he was more intent at this point on illustrating this kind of application of his empirical mapping function which, had it worked, should have simplified the computational tasks involved in the construction of maps significantly.[25]

Paradoxically, the Drosophila group did not call Haldane on either of these issues, or on the issue of whether recombination frequencies could exceed 50% — something which they believed and Haldane did not. I suspect that they felt that the last two issues could not be settled clearly without better data, and weren't going to waste their time arguing about it when it couldn't be settled. It is more puzzling, given that they used it on Castle, that they didn't fault Haldane (or at least his analysis) using Muller's data. I believe that in the intervening two years (between 1920, the time of their last attack on Castle — and his capitulation, and 1922, with the publication of the second edition of *Mechanisms of Mendelian Inheritance*), their views had undergone a more radical methodological shift. This methodological shift may have made Muller's data seem less relevant to them because it suggested an attack at a more fundamental level. For this attack, more accurate empirical studies directed towards a uniform answer, (Muller's, or unpublished studies by Bridges in search of mapping functions — referred to in Morgan, Bridges, and Sturtevant, 1925, pp. 97, 99) and more sophisticated modelling in the spirit of Haldane's (like that of Jennings in 1923) were equally misdirected. The quote which began this paper said it all, and left no room for compromise: *Haldane's view is wrong not because he didn't have the right formula, but because there is no general formula.*

XIII. PARTICULARISTIC MECHANISM AND
THE LIMITATIONS OF GENERAL THEORY

The Morgan school had to wage a series of battles to defend the legitimacy of constructing linear linkage maps and to interpret these maps as indicating relative location of the factors along the chromosome. These battles and the studies they had pursued to determine environmental and internal factors which could affect recombination rates so as to better control for them and eliminate sources of error in constructing the maps had given them knowledge of, and respect for, many particular circumstances which could' cause trouble in the construction of maps. I believe that these particular vagaries of and variations from case to case made a deep impression on them, as well as another feature of their methodology which they probably were not self-conscious of, but which may have had an even deeper effect. They could have quarreled with the detailed form of Haldane's predictive equation, or pointed out that it did not conform with the predictions from their most careful studies, but they chose to quarrel with Haldane on the most general grounds possible — namely the very possibility of giving any general formula whatsoever.

This does not mean that they believed that it was impossible to characterize, understand, or say anything generalizable about the causal mechanisms operating. They were mechanists after all. It does mean, however, that they thought that different particular subsets of the array of causal factors could be operating locally and with different intensity ". . . for each chromosome and for each region of that chromosome (*ibid.*)." Some of these they thought they understood, or were on the way to understanding (for example, Bridges work on non-disjunction and sex-determination, Muller and Weinstein's work on the mechanisms of interference, and Plough's work on the temperature dependence of recombination) and some were still mysterious (Sturtevant's work on "recombination modifiers", and the detailed mechanisms of recombination). Closer study should reveal the causes and their interac-

tions, but until this was available, it would be premature to generalize —
thus ". . . From such careful, detailed region-by-region studies, a more
general statement and formulation should [in time] emerge. Until such
studies are completed, there is nothing to be gained from *a priori*
attempts to formulate the relationship."

This text from *MMH* in 1922 is repeated almost word for word a
year later in Bridges and Morgan (1923, p. 26) — this time without
explicitly mentioning Haldane, though he is clearly in their minds —
and in their bibliography. (In the 1923 text the words "in time" were
added, suggesting even further reservations about achieving a general
account.) This position was further elaborated in substance in Morgan,
Bridges, and Sturtevant, 1925, pp. 97—100, cf. especially p. 99.

This position, which they never stated explicitly, but which seems to
flow fairly robustly from their commentaries on these topics, I shall call
particularistic mechanism, and seems, with various modulations, to
have been held by all of them by the early 1920's. This methodological
viewpoint (1) recognizes the *in principle* generality of causal explanations, but (2) holds that the detailed variability of complex systems
makes it pointless to look for unqualified generalizations that will
predict accurately what will happen in a variety of cases. At the same
time, however, this position would hold (3) that it is possible to determine major causal factors or features or mechanisms that, taken individually, appear in a variety of different — sometimes in all or virtually
all — cases, but appear in different (a) combinations and (b) modulations in different cases, and (4) that close and detailed inspection of any
individual cases would give, subject to technological and experimental
limitations, as much detailed information as desired as to how the
generally applicable kinds of causal factors were interacting in that case.
The first two points are perhaps most clearly stated by Sturtevant in a
long methodological discussion occasioned by six years of study of his
still recalcitrant "recombination modifiers":

... Other things being equal, chromosome sections of equal length will give equal
percentages of crossing-over: but in no case can we be certain that "other things" *are*
equal. The terms "distance" and "percentage of crossing over" have sometimes used
almost as though synonymous and confusion has perhaps resulted: But it has been
recognized from the beginning that different regions might show different frequencies of
crossing over for the same actual length of chromosome. The results presented in this
paper show conclusively that this is the case ... (Sturtevant, 1919, p. 328, *italics in
original.*)

The third and fourth points are more difficult to document. I have never seen them made explicitly by any member of the Drosophila group, though they are both implicit in everything that they do. For the third point, I suggest that the idea of different combinations of causal factors, and of modulations of existing delimited major factors, would both have been natural analogical extrapolations from the genetic systems they studied to the interaction of causes in complex biological systems in general.[26] The idea of different assemblages of different causal factors would have been a natural generalization from the independent assortment and linkage of different non-allelic genes (which were even called "factors" before the commitment to their nature as chromosomal entities was firmly established). Similarly, modulation of action of existing factors would have been a natural extension of the idea of a genetic modifier, used by Muller in 1914 to account for Castle's "hooded rats" case of the supposed "contamination" of a gene in the heterozygous conditions, and used extensively in subsequent explanations by all members of the group. [27]

The fourth point is also constantly reflected in their practice. Thus, the two pages following the above quote are devoted to explaining away three resistant and exceptional cases which appear to compromise the order of the linkage map proposed, each in a different way, using elements of the mechanisms and experimental design supposed to be plausibly operative at the time. After the fact, this may be seen as an ironic mistake. It was used to argue that there has been no shift in the order of genes in the experiments with "recombination modifiers", *which* — given our retrospective knowledge that the "recombination modifiers" were generally inversions — *was probably exactly what did happen!* But this makes the point even more clearly that the "explanation" offered illustrates the methodological commitment to being able to explain the individual variability from case to case on a piecemeal basis. (See also Wimsatt, 1976b, where this kind of model of mechanistic-reductive explanation is presented and defended in detail. Related views have since been presented by Cartwright 1983, Kitcher 1984, Sarkar 1989, 1992, Waters 1990, Kincaid 1990, Bechtel and Richardson 1992, and Burian, forthcoming.)

XIV. THE EMPIRICAL PROOF OF THE PUDDING

What had happened by the early 1920's to change their minds? Two more things that I have not mentioned so far. The first was the acquisition of additional data. This new data was used to update their maps, and also appeared to have generated the second factor, a significant increasing reflection on their methodology — or at least they wrote more of it down. (See Bridges and Morgan, 1919, 1923, and Morgan, Bridges and Sturtevant, 1925). This data for the first time showed unequivocally that *interference varied from region to region within the chromosome* (most obviously in the third chromosome, less so in the second, and comparatively little in the X-chromosome). The *inter*-chromosomal variation in *intra*-chromosomal variability made their second point. Ironically, the X-chromosome was least variable along its length, probably because the spindle-attachment point was at the end, rather than in the middle. Haldane fit his models to the first case they had significant data for — which also happened to be the case where they worked best. (Compare the relative non-uniformity of correction curves for the third chromosome (Figure 10 below) with the much greater uniformity of correction curves for the X-chromosome (Figure 11 below).) The empirical basis of their differences with Haldane could not be rendered visible until their presentation as "correction curves" which indicated the deviations from linearity in the relation between recombination frequency and map distance for different map distances from various genes. This was done for the first time in 1923 in Bridges and Morgan's monograph on the third (and longest) chromosome of *Drosophila melanogaster*. It was the "satisfactory representation of such relationships" promised in *The Mechanisms of Mendelian Heredity* in the quote that began this paper.

In Figures 10 and 11, labelled curves indicate the deviation from linearity, or the difference between map distance and observed recombination frequency, as a function of location in the map (or actually map distance from one end.) They are directly convertible into curves like Haldane's (Figures 8 or 9 above) by plotting the X-coordinate minus the deviation as the Y-coordinate.[28] Figure 10 makes clear what the Morgan school has maintained all along — that no single mapping function will do — that the necessary mapping function varies from region to region in the same chromosome. Figure 11 extends this to different chromosomes. The same mapping function throughout the whole chromosome would entail that the correction curves emanating from different parts of the chromosome would remain parallel, so they would be superimposeable via translation along the map axis, and they clearly are not.[29] Their discussion is unambiguous on this point:

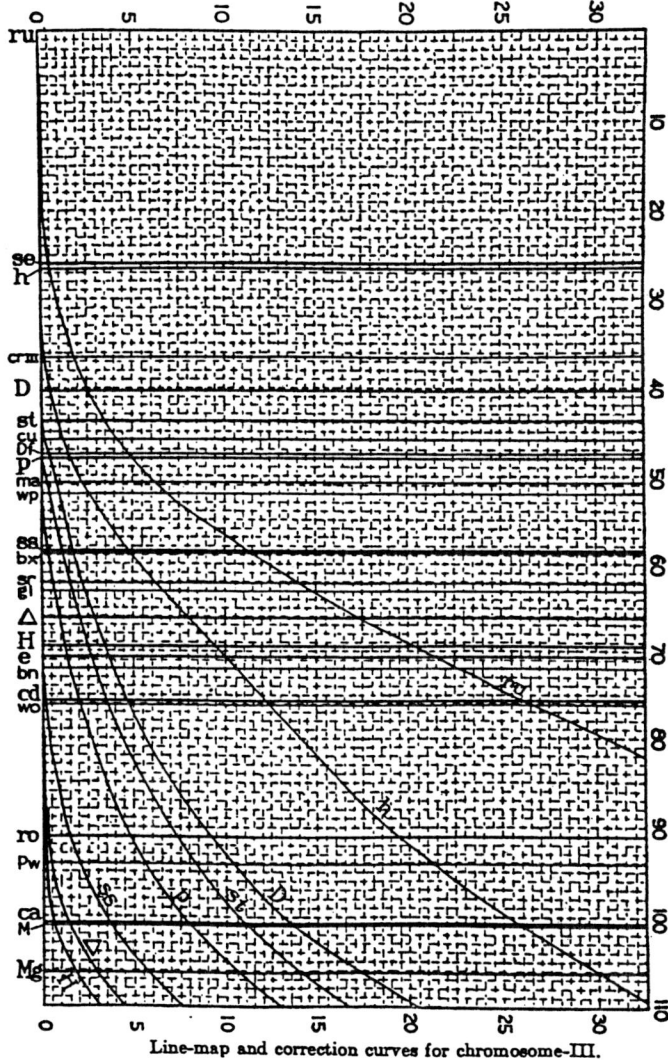

Line-map and correction curves for chromosome-III.

Fig. 10. Correction curves for the third chromosome, giving differences between observed recombination frequency and map distance from different genes as a function of location on the line map. If a single map function exists, these curves should be superimposeable through lateral shifts along the line map. From Bridges and Morgan, 1923, fig. 1, p. 12. Reproduced courtesy of the Carnegie Institution of Washington.

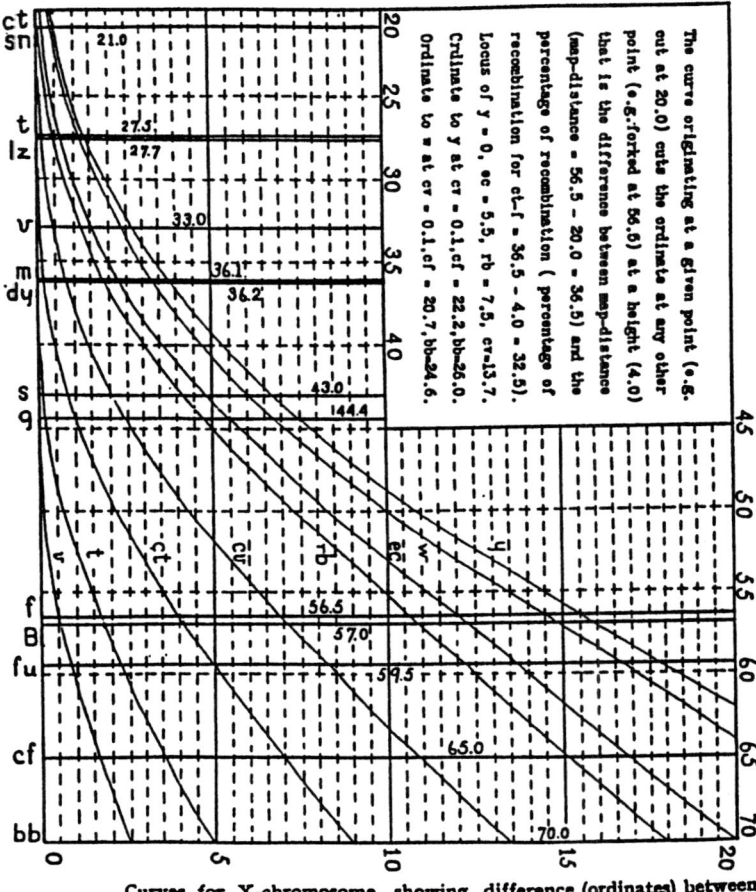

Curves for X-chromosome, showing difference (ordinates) between map-distance (abscissae) and percentage of recombination.

Fig. 11. Correction map for the X-chromosome, originally published as Figure 43, p. 103, of Morgan, Bridges, and Sturtevant, 1925. Although based on "early" data apparently collected and graphed by Bridges some time before 1919 (cf. 1925, p. 99), this figure first appears in 1925, two years after the corresponding correction map for the later studied 3rd chromosome. They may not have hit upon this way of representing the data before 1923, or the delay may have reflected the hypothesized change in their methodological orientation described in the text — to a view which would have been better represented by the irregularities of the 3rd chromosome map. Ironically, this map, once it appeared, was much more widely reproduced in texts than that of the 3rd chromosome, possibly because the X-chromosome was more commonly emphasized.

... It should be noted especially that the curves originating at the various loci are specifically and rather widely different from one another. This becomes strikingly apparent from a comparison of ordinates at equal distances from the origins of, for example, the *p* and the *ss* curves, or the *ru* and the *D* curves. But these differences are themselves in an orderly system, which is connected with the regional differentiation of the chromosome. The length of section, measured in units of map distance, within which a double crossover can occur, is shortest in the region about pink [*p*] and longest in the region about white-ocelli [*wo*], and has a characteristic value for each region of the chromosome, except that the values are probably symmetrical about a point lying near warped. Accordingly, all of the curves having their origins to the left of pink make decided rises as they pass this mid region, those originating near-by at the left have rises early in their course, while those originating further to the left rise progressively later in their courses. To attempt to express by a single curve or formula this relation between map-distance and recombination, would only obscure the real situation, and the result would be as meaningless as a map based upon all the crossing-over data, irrespective of known modifiers of crossing over. (Morgan and Bridges, 1923, p. 26.)

They then go on with the passage quoted earlier about capturing the relationship on a region by region basis, and express the hope that a more general formulation should emerge from this. I suspect that this "more general formulation" would not — for them — have been algebraic however. Rather, their remarks in several places suggest that *they use the very modulation of the relationship as a clue to the existence of additional mechanical factors* (such as the role of the spindle attachment point, discussed below) *which, while being quite individual in how they apply from case to case, nonetheless point the way to a more complete causal account of what is going on.*

This is a data-driven analogue to the use of incomplete (analytical) models in Wimsatt, 1987. The point there was that a simple analytical model could be used to estimate the magnitude of effects of variables left out of the model by looking at the differences between the model's predictions and experimental realities. In the above example of the correction curves for the third chromosome, they do not have any analytical models for the deviations from linearity with increasing distance, but they do have empirically derived curves, and the qualitative changes in the behavior of these curves (perhaps with the converse of "same cause, same effect") suggest the existence of changing causal factors modulating recombination along the physical chromosome and linkage map. This is a very important reasoning pattern — one which, if I am right, is central to their methodology as particularistic mechanists. I will return to it in the next section.

It is important to see that *systematic irregularities in interference which were a function not just of the distance between crossovers, but of*

their location, undercut the generality of Muller's work (though importantly, not its usefulness for analyzing mechanisms) *just as surely as that
of Haldane*. The existence of a coincidence map that specified frequency of double crossovers (or even triples, quadruples, etc. . . .) *just*
as a function of distance would simply entail and be convertible (for
practical purposes, using some curve-fitting procedure) into a mapping
function. Muller's data-based method of constructing a coincidence
curve might have allowed mapping functions so pathologically irregular
that Haldane would never have dreamt of them, but his method nonetheless basically presupposed that coincidence could be specified as a
function of distance. As it became clearer that coincidence varied with
the location of the segments as well as the distance between them, the
motivations for the extra work of doing coincidence curves must have
vanished. This awareness must have existed by 1923, since it is entailed
by the different and location dependent correction curves for the third
chromosome.

If Haldane's project was one that they had, in some sense, attempted
and found wanting, it would help to explain their opposition. One
cannot date their fall from this state of grace to venal particularity
exactly. Nonetheless, given what else was happening by 1919, I think
that they were well on their way to systematic doubt by then — if not
already over the edge. Gowen's careful biometrical study showing
substantial linkage variations along the third chromosome was sub-
mitted in November of 1918; Morgan (1919, p. 130), reports that
Bridges has found a modal length between crossovers of about 15 in
the middle of the second chromosome, but about 30 nearer the ends,
and notes Weinstein's figure of 46 for a modal distance in the
X-chromosome; and the use of a correction formula by Bridges and
Morgan in 1919, (p. 299), is not general at all, but tentative and local in
the extreme. If I am right, they were probably already committed
sceptics of general theory in this area by the time they could have read
Haldane's paper. The evidence which leads in this direction demands
further development, but that is a topic for another time.

XV. DISPUTES AMONG ALLIES?
CONCLUSIONS ON TOP-DOWN VS. BOTTOM-UP THEORIZING

So, it is tempting to ask, "Who was right?". Even if we assume that this
question *has* an answer, it would appear that, in the area of linkage

mapping, it hasn't yet been decided. In a recent review (1979) Joseph Felsenstein clearly recognizes the validity (and correspondingly, the limitations) of both kinds of approaches. He sees the advantages of his approach, in which he derives a new general mapping function, in simplicity, useful formal properties, the derivability of other important models — including Haldane's and Kosambi's — as special cases, and predictive and calculational benefits. But he sees its limitations too, and points them out with a wry sense of humor:

Our mapping function is purely phenomenological, derived without any underlying model of the recombination process. Its parameter, K, is of only limited meaning biologically, giving us little more insight than the raw data. There are a number of papers in the genetic literature in which specific mapping functions are derived which are predicted by particular models of the recombination process. While these models have the advantage of precision, they run the risk of being made irrelevant by advances in our understanding of the recombination process. In this respect, the very lack of precision of the present phenomenological approach makes it practically invulnerable to disproof. (Felsenstein, 1979, pp. 773—4.)   .

It is tempting to ally Felsenstein with Haldane, and the Morgan school with those unnamed moderns who base mapping functions on particular models of the recombination process. But this is too simple: given the level of resolution of the time, Haldane's models (as represented by equation (5)) had a causal interpretation too — only his last predictive equation was phenomenological. And, although the discussions and analyses of the Drosophila group were dominated by a rich interacting fabric of possible causal interactions which they wished to disentangle, their final method of producing the best approximation to the "standard map" was paradigmatic phenomenology and acausal numerical approximation. The causes were lurking in the background to be sure, in the explanations for why the correction-maps had to be so variable and complex, but the method of final map construction — or perhaps it is fairer to say, of final map correction, was (as I will show on another occasion) unabashedly phenomenological. And they didn't produce mapping functions. So both of them had causal motivations and rationales for their models, and both of them indulged in pure phenomenology when necessary — yet clearly they differed, so the dispute is more complicated than it appears.

Where they differed was in the amount of generality they sought, and the amount of particular detail they felt was necessary for an adequate account of the phenomena. This seems to have been a matter of mutually reinforcing experience, probably taste, and the purposes of the theory which they developed from and for the phenomena. One can say that Haldane had a taste and a substantial ability for abstract and generalizeable mathematical theory, and that members of the Drosophila group seemed to relish causal detail, and to generalize, where it seemed called for, in terms of causal factors which could vary in their detailed mode of interaction and intensity rather than in terms of algebraic formulae. Perhaps revealingly, the only things they schematized were their standard crosses, whose diagrammatic representations (e.g., *MMH*, 1915, Fig. 12, p. 23 for a 2 factor "Punnett square" with chromosomes, and Fig. 23, p. 55 for a 2-factor sex-linked cross, for two common kinds of diagrams) became both emblematic and paradigmatic for their science and modular tools in the construction of complex experimental designs. These were flow charts or procedures, each having inputs and outputs, which could be drawn from like a library of routines and put together like a computer program to assemble a complex experiment. (I owe these parallels between experimental design and the design of computer programs, to my frequent collaborator, Jeffrey Schank, who discusses the three-way parallels between mechanisms, experiments, and programs, in much greater depth in his 1991.)

I think one can say fairly that the Drosophilists were closer to the data, in that they were all immersed together in an ongoing research project which spanned many years, produced an enormous body of data, which by its character, begged to be treated systematically, but also given the complexity of modes of interaction and of experimental control they were discovering and analyzing, had to recognize the individuality of each experimental design, each chromosome, each trait, and each combination of traits. By contrast, Haldane's discussion of 1919 seemed to leave no room for anything less than universality. He suggested that his mapping function or predictive equation be tried out in different places as a test (a test in which he saw mostly positive qualitative confirmation of his theory), but there is no suggestion that if it worked in some places and not in others, that would have been anything but a disconfirmation. There was a total lack of consideration of the significance of inter-chromosomal or interorganismal variability by Haldane. But why were these differences so important — I am tempted to say, to a man — to the Drosophila group?

The key was not just that they noticed the differences and saw in them possible sources of error in generalization, and didn't want to paper over them (prematurely.) *Their methodology was such that they used these differences.* (Ian Hacking's version of entity realism (Hacking, 1983) is based in experimental usage: If you can spray electrons, (e.g., in a cathode ray tube to make oscilloscopes, televisions, and computer screens) they exist. Perhaps the point here is similar: that *if you use the differences, you can't ignore them, and find it harder to generalize over them.*) That they used these differences — ingeniously — is beyond dispute, and I will document this in much greater detail on another occasion. *These variations were used in explanations, to get a handle on how to criticize competing theories or to refine or modify theories which they wished to hold, and especially — wherever they could — as incorporated parts of laboratory designs, procedures, and apparatus.*

Variations between the X-chromosome and the second and third chromosomes were related to differences in the attachment points of the spindle fibers. Variations within the third chromosome were related to its V-shaped structure and the location of the spindle attachment point. The peculiar linkage behavior and sensitivity of the mutant Purple (*pr*) to linkage disturbances with temperature and recombination modifiers was attributed to its location near the center of the chromosome and the spindle attachment point. The high recombination rate (which gave statistically accurate data) and general sensitivity to disturbances of the middle of the third chromosome led to its use as a "litmus test" for the presence of other disturbing factors in specific experiments. The Morgan school didn't just glory in the differences — they used them. Like a mountain climber, they used any imperfections in the regular face of the phenomena to guide their ascent to a higher (and deeper) understanding of the mechanisms, but, unlike the mountain climber, they could also bring along any imperfections they found as tools to help in locating other means of ascent and other chinks in other seemingly regular and important faces.

This is as true for false generalizations as it is for quirky exceptional pheonomena. As pointed out above in section 2 above, Trow's rule was honored by the Morgan school — almost daily — but always in the breach. It was an ideal template against which one could see the order that was real. Since it was deviations from Trow's rule that indicated interference, Trow's rule was invoked or used every time they calculated coincidence. What made the Drosophila group's use of Trow's rule different from the perspective I have advocated in the past

(Wimsatt, 1987) was the tacit assumption there that false models lead to truer theories (true) and that these theories would at least aim for generality (which these examples show to be ambiguous or false — at least as a general claim.) I will return to this shortly.

I have focussed on the few years immediately preceding and succeeding the publication of Haldane's 1919 paper, and on issues connected with linkage mapping, but the whole tissue of the discipline, before and after this period supports this kind of vision of the use of anomalies as tools. Mutations went beyond Mendel's laws, and linkage contradicted his second law, yet the search for and later the generation of mutants and the analysis of linkage relations were so central to their methodology as to be almost constituitive of it. Manipulation of non-disjunction, producing extra or missing chromosomes (an abnormality if there ever was one) gave Bridges' "final proof" of the chromosome theory, and generated the raw materials for a fuller theory of sex-determination. (And recall that Boveri's classic dispermic fertilization experiments of a few years earlier (1902, 1907) used a naturally rare anomaly (which produced extra or missing chromosomes in spades!) as part of an experimental design which gave the most robust and convincing evidence before that of Morgan's group that Mendel's factors *were* located on chromosomes.) The discovery that there was no crossing over in Drosophila males was not so much an embarrassment as another fact to be used in designing experimental techniques. After the discovery of the giant salivary gland chromosomes, with their visible banding patterns, deletion mapping (in which parts of the chromosomes were lost) became the tool for correlating the locations of genes in the linkage map with the locations of bands in the chromosomes. The list is virtually endless, and it is hard to conceive of a biological experiment today which does not make use of tools (often multiple tools) which are co-opted anomalies. It is an old saw among biologists and explorers of any unknown complex systems that the best way to learn how a mechanism works is to study how it breaks down. Doctors, engineers, programmers and auto-mechanics can all appreciate the truth of this maxim. Induced breakdowns — the heart of experiments — involve deliberate manipulations to disturb regularities of behavior. The exception may not prove the rule, but it may be crucial to understanding how or why it works.

Even if all of this is so, wouldn't the members of the Drosophila group think that sooner or later they would have all of the causal factors and could then construct an exceptionless general theory? I doubt that any of them were ever asked the question in this way, and who knows what they would think, but it seems to me that there are two plausible answers. First, if the end of science is seen as knowing all of the causal forces, but they can be instantiated in different ways in an indefinitely (not infinitely!) large variety of different mechanisms, then knowing a general theory in this sense does not give you an exceptionless general equation. Second, if your procedure focusses on behavior at the margin — always using the latest imperfections to understand the limitations in the causal generalizations you have so far, and to detect new ones to be flawed in their turn, rather than focussing on the unattained and idealized endpoint — you may not generalize to look for an exceptionless generalization at the end. You may instead generalize the cyclic process you have followed, and always expect the next round of imperfections, and look forward to the toehold it will give you on the ever more finely detailed mechanisms. This was the difference in their perspectives, and I think it would be hard for any experimentalist to totally escape the latter viewpoint. Was God a general theorist, or a watchmaker for whom it was escapements all the way down? For a co-opter of anomalies, could there be any other answer?

ACKNOWLEDGEMENTS

None of this would have happened without the opportunity to teach some history of genetics provided by my Biology common core course — Genetics in an Evolutionary Perspective, a preoccupation now of 20 years standing. That course provided the puzzle — since the Haldane Mapping Function perfectly captures the effects of multiple crossing over, how come the Morgan school didn't use it? (This was a puzzle perhaps only to me, because I had initially assumed that they did!). At a later stage, when a cursory inspection had led me to wonder whether the Morgan school had ever even cited Haldane's 1919 paper, Sahotra Sarkar, who knew of my interest, found the crucial reference to Haldane in the 2nd edition of *The Mechanism of Mendelian Heredity*. Their criticisms there indicated the directions for a much more pointed search, which put this project on track. Jim Crow has been an invaluable source of advice on more recent theories of linkage and key intermediate historical articles which made backtracking much more productive. Both have been frequent sounding boards, both for moulding various ideas expressed here, and also for helping to cull some of the ones which did not survive. Gar Allen, Dick Burian, Elof Carlson, Lindley Darden, and Will Provine have provided useful bibliographic advice and informed assessments to a variety of "judgement" questions which, I hope, have pointed some of my interpretations in the right direction. Jeff Schank gave it a close reading for intelligibility and consistency, and has been particularly helpful as a sounding board on the attitudes towards and uses of laboratory procedures. I have had the best possible sources of advice. Any remaining failures are in spite of the most effective help! Portions of this work were done with support from the National Science Foundation (Grant — SES8807869) through the History and Philosophy of Science Program, Ronald J. Overman, director. I thank them for their continued support.

The University of Chicago

NOTES

[1] This is less true of Boveri's work than of Sutton's, since Boveri's work with dispermic fertilization (Boveri, 1902, 1907) showed major developmental pathologies if the right combination of chromosomes were not inherited, and had some implications for the time of onset of gene action. Nonetheless, Boveri's work did not show, any more than Sutton's did, how the factors accomplished their developmental role and how differentiation of cells which received the same complement of factors was to be explained.
[2] Without further information, it also does not tell which end of the map corresponds to which end of the chromosome.
[3] The Trow formula is not equivalent to the Haldane mapping function, but it is rather clearly part of the motivation for the Haldane mapping function. The main differences are that the Trow formula has the form of an expression giving the predicted observed recombination frequency as a function of the observed recombination frequencies in two segments that compose it, rather than recombination frequency as a function of map distance. The second difference is that the Trow formula corrects for the effects of double cross-overs, but not for triples or higher, whereas the Haldane mapping function generalizes this and corrects for the expected frequency of all higher order crossovers.
[4] Equations are numbered differently here from in Haldane's paper because he does not number all of the equations he discusses. To reduce confusion, the numbering in quotes from that paper is changed to reflect that used here.
[5] He actually also did and reports on extensive experiments following multiple factors simultaneously on the second (10 factors) and third chromosomes (7 factors). In fact, to reduce the number of lines that had to be manipulated he later combined and worked with the multiple mutant X and second chromosomes in a single strain, which was then heterozygous for 22 mutants!

[6] The number of classes that would need to be considered rises exponentially with the number of mutants being followed, so this method could only be used to correct for viability effects for two or possibly three mutants at a time, as Muller himself apparently recognized.

[7] See for example Weinstein, 1918, for further theoretical and experimental development of the method, and Sinnott and Dunn (and later Dobzhansky), where it is missing from the first edition of 1925, but discussed extensively in 1932 and subsequent editions; Sturtevant and Beadle, 1939; and Strickberger, 1968; for three textbook presentations spanning four decades. Interestingly, Snyder (1935) breaks with national lines to follow Haldane 1919, in his mode of presentation — (without explicit citation, but compare p. 158), refers to interference, but not coincidence — a practice apparently more common among English writers.

[8] There is an ambiguity implicit in characterizations of interference — between the physical process of interference and statistical measures of it. Thus it is possible for (positive) physical interference to produce negative (statistical) interference — indicating greater than random probability of crossing over at a given distance if the mechanism of crossing over indicates a modal distance between crossovers.

[9] Indeed, without changing the form of his procedure, Haldane *could not have fitted* curves that asymptoted with increasing map distance to a recombination frequency of greater than 50%, or curves whose form indicated a coincidence of greater than 1 (or *p* of greater than 2). This is because he used assumptions that the correct answer could be found at or in-between his two limiting cases, both when he was model-building and when he was constructing his best-fit curve, and these kinds of curves would, in different ways, fall outside of this range.

[10] Indeed, the situation is even worse. Muller notes (1916—IV, p. 421) that the map distances he gets are all approximately 6/7ths of the lengths from the "composite maps" (i.e., the maps derived from combining the results of many different experiments commonly used by others — e.g., as found in Morgan and Bridges, 1916, or Morgan, Sturtevant, and Bridges, 1925). Muller's estimates are used in this section, but these differences must be kept in mind when locating factors, estimating distances, or making comparisons between his results and those of others — particularly Weinstein, 1918. Strangely, Muller attributes this difference to systematically lower crossover rates among "a few females", but chooses to include their data, rather than to throw it out.

[11] I have in this exposition added some details which he left unspecified, but which are required or implicit in his presentation.

[12] This may not have been accidental. Thus, their evaluation maps (Bridges and Morgan, 1919, p. 299, Bridges and Morgan, 1923, p. 27), used in planning experiments both used an interval of 2 units, and they may have regarded this as a plausible "maximum error" in their map locations — indeed, it may have been very close to their maximum cumulative error even for genes very far apart whose distance they would not have measured directly. Certainly most revisions in the map locations of mutations were much smaller than this, and distance changes this large or larger were used as grounds for positing another "system" for linkage data, suggesting a non-standard map produced by translocations or other unknown causes. (*Ibid*, p. 20.) Estimates of the maximum length of chromosomes changed by more than this, but only through the discovery of factors which fell outside of the previously known ones.

[13] Since the graph shows both changes in direction of the curve at values other than integral multiples of 2, and successive changes in direction with separations of less than 2, unless he was very sloppy in doing the graph, Muller could not have constructed it using a Δd of 2 with all intervals aligned. So the aggregate curve must have been produced either by using a smaller Δd, for at least some of the curves, or by individually choosing for each double recombination event, the starting point for the increment so that it provided a best fit to the two intervals and their separation for that particular case of double crossover. A common grid at integral multiples of 2 would provide a pretty poor fit, especially for the smaller intervals and separations. Using a smaller Δd for some of the curves would have increased the computation for those curves, but Δd's of different magnitudes could easily be used for different cases, if the larger ones were integral multiples of the smallest, and would have been less confusing than using asynchronous intervals for the different doubles. He may have done both. In constructing the graph, he also appears to have connected lines drawn from the midpoints or edges of the intervals (which the pointiness of the peaks suggests), rather than insisting on a very fine-grained bar graph. This is done in the worked example graph of Figure 5, which (by contrast with Figure 6) *was* done with common synchronized Δd's of 2.

[14] For its conservatism, Muller's procedure was thus well designed to demonstrate the existence of interference to a sceptic, though perhaps less well designed to measure it. He could perhaps have done better by trying different hypothesized interference distances (or distributions of crossover separation distances) to see which produced the best fit with the data (according to some criterion), thought this would have increased what must have already been an onerous computational burden.

[15] Noting that Muller's distances are 6/7 of the standard ones, this would promote his 8 to 9.3, and still leave a difference of 4.2 "standard" map units at the lower end and much more at the upper end (his 36 now becomes 42 "standard" units) which must be allowed for in the upper end biassing described in the next paragraph. Note that Muller's method (in Figures 6 and 7) assigns small but finite probabilities to second crossovers at distances down to 0, even though none had been found at distances less than 13.5.

[16] It was improper only to the extent that the biasing effects of this procedure were not recognized, but I have seen no explicit recognition of them, and other claims that suggest it was not recognized. Thus Haldane says (1919, p. 300) that he has seen no evidence for peaks at integral multiples of a modal distance, although both Muller (1916) and Weinstein (1918) do provide evidence for coincidence values slightly above 1 at separations that would be plausible modal distances, declines at greater distances, and Weinstein points out that the chromosome might not be long enough to have higher order peaks — which seems to be the case for the X-chromosome.

[17] Muller's concern to make the construction of the theoretical curve parallel with that used to analyze the data seems well founded, since the exact form of the composite curve should vary with different partitionings of the chromosome into intervals. Unfortunately I do not see how any combination of frequencies with these different length intervals and this method of construction will give the curve represented by the dotted line in Figure 6. I find its flat top and linear decline from that top with increasing distance particularly suspicious. I plan to redo the calculation to see what assumptions are required to get his results, but I have not yet had the opportunity.

[18] Muller appears to be too conservative here in two respects: As noted above, his procedure for treating the data was conservative with respect to showing a modal peak as well as with respect to showing interference at closer distances. His estimate of the height of the peak (at about 1.05) was clearly a lower bound, and almost certainly an underestimate from his data. Secondly, the curve *is* in fact roughly symmetric around the peak at 35 centimorgans from about 25 to 45 centimorgans. What really disturbs symmetry is the apparent second rise or peak at about 46 or 47, and if the data is increasingly unreliable with longer distances as he says, perhaps *these* points should have been discarded rather than all of the points above 35, preserving the symmetry. He may have not wanted to treat the data in this way to avoid charges of favoritism towards his own hypothesis, or he may have had other reasons for marking the uncertainty as starting at 35 centimorgans, but if so, he does not make them explicit.

[19] I think it is at least equally likely that the idea originated with Muller, who discusses it in detail in his dissertation in 1916, and had obviously been working with it for some time before that, though it is also discussed (without all of the implications) in Bridges, 1915. See discussion below, and also Carlson's biography of Muller (Carlson, 1981) for discussion of Muller's priority disputes.

[20] Despite this ambiguity, I think it is clear (because it is so clearly overdetermined) that Haldane did intend a contrast between his causal-realistic models and his instrumental-predictive one. This kind of case, where one and the same theorist provides different but predictively virtually indistinguishable models, and clearly does regard them as different kinds of beasts should produce an interesting challenge to those, like van Fraasen 1980, who wish to give an instrumentalist-predictive account of all theory. Such cases are worth closer attention by philosophers of science.

[21] According to informal gossip from several sources, Haldane also had a somewhat casual attitude towards checking his work, which led to many errors in his printed articles. In the process of entering and recalculating his data to do the graph for Figure 9, I found further confirmation of this in the form of 3 independent errors in table I, which because of serial dependencies in calculations, generated 2 more, and an unrelated graphing error. The latter is in the location of the point closest to being on a straight line between the labels *a* and *c* in his graph (Figure 8) and can be detected by looking for the corresponding point in Figure 9 below. Although none of these errors would have made a significant difference to any of Haldane's results, they would have generated some amusement if detected by the Drosophila group, and surely would have contributed to Morgan's well-known disdain for theoreticians. (Allen, 1979, and personal conversation). In any case, five errors out of 108 calculated values is a bit much. Thus, from a survey of 1300 data papers in the social sciences, noted social psychologist Robert Rosenthal reports an average error rate of about 1% (lecture, EVIST conference, Cazenovia, 1981).

[22] More crucial than requiring a single experiment (however that might be individuated) is the appropriate controls on the experiments done, though doing it all in a single experiment could make it easier to control for unknown effects. From data available to them, the factors controlled would have to include that all of the experiments in question, if more than one, be done on the same strain of flies of the same age and at the same temperature (Plough, 1917). Use of the same strain of fly should control both for genotypic viability effects and chromosomal rearrangements, like the "recombination modifiers" discussed by Muller (1916, 1920) and Sturtevant (1915,

1917, 1919) and finally explained as inversions by Sturtevant in 1921 or 1926. Standing procedure already controlled for the age of the fly (Bridges, 1915).

[23] The range of data in the two sets is not the same. Haldane explicitly did not include lower distance values from Morgan and Bridges' data because they would be more nearly additive and would not be useful in discriminating between the various mapping functions (pp. 305—6). The data left out were also more well behaved — less noisy, just as Muller's are in this lower range. Since Haldane's data both starts higher and extends higher, it might for this and other reasons be expected to show larger absolute variation, but this is not completely true. Higher recombination frequencies mean that the number of individuals in different classes will be more evenly distributed, reducing the particularly strong effects of sampling error when some classes are very rare. For this reason the later development of theory focussed on questions like: "What is the optimal separation between genes to use in constructing maps in order to minimize error?" See, e.g., Haldane, 1931.

[24] As bad as the comparison is, it actually gives Haldane the benefit of the doubt. If one takes into account Muller's claim that his measured frequencies and map distances are 6/7ths of the real ones (see notes 10 and 15 above), and adjusts for it to give probable true distances (as he anomalously did not), the proper correction would be to multiply x and y values by 7/6. If we made this correction here, Muller's curve would diverge from the diagonal later, and would be above Haldane's "best fit" curve by an even larger amount. This adjustment would further exacerbate how bad Haldane's "best fit" curve performs for Muller's data. It wouldn't have affected their predictions of the length of the X-chromosome, since Muller didn't use any factors too close to either end, and they didn't use mapping functions to estimate this anyway.

[25] He proposes other sources of uncertainty not discussed here. Haldane mentions the cumulative uncertainty that might come from constructing the map from a larger number of short distances, but this would require either a systematic deviation which he has given no reason to expect or a probabilistically unlikely coincidence. He suggests also that undetected multiple crossing over in the longer distances could yield the shorter map. This would have seemed unlikely to the Drosophila group at the time, since the longest map separation involved was 15 units, while the closest crossovers observed for the X-chromosome to that time was 13.5 (Weinstein, 1918). With this difference, undetected double crossovers could not have been frequent enough to explain the substantial difference in estimated map distances.

[26] This is a point which deserve substantial further study. Historically, a number of techniques of statistical analysis (including for example both the analysis of variance and the method of path coefficients) originated through generalizations of techniques constructed first for the analysis of special cases in genetics. It is likely that the same has happened for the general conceptualization of causal systems. There is a strong suggestion that this happened for Sewall Wright for his method of path coefficients. (See, e.g., Provine, 1986.)

[27] Both of these in turn are examples of using partial models to factor out a component of variation in order to focus more clearly on what is left, an extremely important pattern in causal modelling — See Wimsatt, 1987. For a detailed case study of the action of modifiers on a character, see also Morgan, 1919b.

[28] This is in fact done by them, on the facing page (p. 13) but is enormously more confusing, since they have also drawn in the diagonal line, $x = y$ for each major locus.

[29] Actually, the eye is misleading — even though they clearly are not parallel, they deviate from being parallel by more than might appear. The difference in map coordinates (Δx — the long axis) in the corresponding curves should remain constant for difference deviation values (Δy — the short axis) which will be true if and only if all of the curves exhibit the same mapping function.) This reminder is not required for the 3rd chromosome map, where the non-linearities are quite obvious, but should be kept in mind when looking at the corrections curves for the X-chromosome, which look far more linear.

REFERENCES

Allen, G. (1979), *Thomas Hunt Morgan: The Man and His Science*, Princeton: Princeton University Press.

Bateson, W. and R. C. Punnett (1911), 'On Gametic Series Involving Reduplication of Certain Terms', *Journal of Genetics* **1**: 293—302.

Bechtel, W. and R. C. Richardson (1992), *Discovering Complexity: Decomposition and Localization as Scientific Research Strategies*, Princeton: Princeton University Press.

Boveri, T. (1902), 'On Multipolar Mitosis as a Means of Analysis of the Cell Nucleus', English translation of 'Über mehrpolige Mitosen àls Mittel zur Analyse des Zellkerns', *Verhandlungen der physikalisch-medizinischen Gesellschaft zu Würzburg* **35**: 67—90, reprinted in B. H. Willier and J. M. Oppenheim, *Foundations of Experimental Embryology*, 2nd ed., New York: MacMillan, 1974, pp. 74—97.

Boveri, T. (1907), 'Die Entwicklung dispermer Seeigel-Eier. Ein Beitrag zur Befruchtungslehre und zur Theorie des Kerns', *Zellen-Studien*, Heft 6, Jena: Verlag von Gustav Fischer, pp. 1—292.

Bridges, C. B. (1915), 'A Linkage Variation in Drosophila', *Journal of Experimental Zoölogy* **19**, #1 (July): 1—21.

Bridges, C. B. (1935), 'Salivary Chromosome Maps — with a Key to the Banding of the Chromosomes of *Drosophila melanogaster*', *Journal of Heredity* **26**: 60—64.

Bridges, C. B. and T. H. Morgan (1919), The Second Chromosome Group of Mutant Characters, Carnegie Institute of Washington, Publication No. 278, Part II, pp. 123—304.

Bridges, C. B. and T. H. Morgan (1923), The Third Chromosome Group of Mutant Characters in *Drosophila melanogaster*, Carnegie Institute of Washington, Publication No. 327.

Burian, R. (1991), 'Underappreciated Pathways to Molecular Genetics, or How to Accomplish a Synthesis in Biology', talk at Chicago, February, 1991, and forthcoming as a paper in a symposium on molecular biology edited by S. Sarkar, thru Boston Studies in the Philosophy of Science.

Carlson, E. A. (1967), *The Gene: A Critical History*, Philadelphia: W. B. Saunders, reprinted by the University of Iowa Press.

Carlson, E. A. (1981), *Genes, Radiation and Society: The Life and Work of H. J. Muller*, Ithaca: Cornell University Press.

Cartwright, N. (1983), *How the Laws of Physics Lie*, London: Oxford University Press.

Castle, W. E. (1919a), 'Is the Arrangement of the Genes in the Chromosome Linear?', *Proceedings of the National Academy of Science* **5**: 25—32.

Castle, W. E. (1919b), 'Are Genes Linear or Non-Linear in Arrangement?', *Proceedings of the National Academy of Science* **5**: 500—506.

Coleman, W. S. (1970), 'Bateson and Chromosomes: Conservative Thought in Science', *Centaurus* **15**: 228—314.

Crow, J. F. (1987), 'Muller, Dobzhansky, and Overdominance', *Journal of the History of Biology* **20**, # 3(Fall): 351—380.

Darden, L. (1991), *Theory Change In Science: Strategies from Mendelian Genetics*, New York: Oxford University Press.

Felsenstein, J. (1979), 'A Mathematically Tractable Family of Genetic Mapping Functions with Different Amounts of Interference', *Genetics* **91**: 769—775.

Fisher, R. A. (1949), *The Design of Experiments*, 3rd ed., London: Oliver and Boyd.

Gowen, John, W. (1919), 'A Biometrical Study of Crossing Over: On the Mechanism of Crossing Over in the Third Chromosome of *Drosophila melanogaster*', *Genetics* **4**: 205—249.

Hacking, Ian (1983), *Representing and Intervening*, London: Cambridge.

Haldane, J. B. S. (1919a), 'The Probable Errors of Calculated Linkage Values, and the Most Accurate Method of Determining Gametic from Certain Zygotic Series', *Journal of Genetics* **8**: 291—298.

Haldane, J. B. S. (1919b), 'The Combination of Linkage Values, and the Calculation of Distance between the Loci of Linked Factors', *Journal of Genetics* **8**: 299—309.

Haldane, J. B. S. (1920), 'Note on a Case of Linkage in Paratettix', *Journal of Genetics* **10**: 47—51.

Haldane, J. B. S. (1927), 'On Being the Right Size', in his *Possible Worlds and Other Essays*, London: Chatto and Windus; also reprinted in E. Nagel and Newman, eds., *The World of Mathematics*, New York: Simon and Schuster, 1954.

Haldane, J. B. S. (1931), 'The Cytological Basis of Genetical Interference', *Cytologia* **3**: 54—65.

Janssens, F. A. (1909), 'Spermatogenese dans les Batraciens. V. La Theorie de la chiasmatypie. Nouvelles interpretation les cineses de maturation', *La Cellule* **25**: 387—411.

Jennings, H. S. (1923), 'The Numerical Relations in the Crossing Over of the Genes, with a Critical Examination of the Theory that the Genes Are Arranged in a Linear Series', *Genetics* **8**: 393—457.

Kincaid, H. (1990) 'Molecular Biology and the Unity of Science', *Philosophy of Science* **57**: 575—593.

Kitcher, P. (1984), '1953 and All That. A Tale of Two Sciences', *The Philosophical Review* **93**: 335—73.

Kosambi, D. D. (1944), 'The Estimation of Map Distances from Recombination Values', *Annals of Eugenics* **12**: 172—176.

Lederman, M. (1989), 'Genes on Chromosomes: The Conversion of Thomas Hunt Morgan', *Journal of the History of Biology* **22**(1): 163—176.

Moore, John A. (1972), *Heredity and Development*, 2nd ed., Oxford: Oxford University Press.

Morgan, T. H. (1909), 'What Are "Factors" in Mendelian Explanations?', *Proceedings of the American Breeder's Association* **5**: 365—368.

Morgan, T. H. (1910a), 'Chromosomes and Heredity', *American Naturalist* **44**: 449—496.

Morgan, T. H. (1910b), 'Sex-limited Inheritance in *Drosophila*', *Science* 32: 120—122.

Morgan, T. H. (1911a), 'Random Segregation versus Coupling in Mendelian Inheritance', *Science* 34: 384.

Morgan, T. H. (1911b), 'Chromosomes and Associative Inheritance', *Science* 34: 636—638.

Morgan, T. H. (1919a), *The Physical Basis of Heredity*, Philadelphia: J. B. Lippincott.

Morgan, T. H. (1919b), 'A Demonstration of Genes Modifying the Character "Notch"', *Carnegie Institute of Washington* Publication # 278, Part IV, pp. 343—388.

Morgan, T. H. and C. B. Bridges (1916), 'Sex-Linked Inheritance in *Drosophila*', Carnegie Institute of Washington, # 237, pp. 1—92.

Morgan, T. H., C. B. Bridges, and A. H. Sturtevant (1925), 'The Genetics of *Drosophila*', *Bibliographia Genetica* 2: 1—262.

Morgan, T. H., A. H. Sturtevant, H. J. Muller, and C. B. Bridges (1915, 1922), *The Mechanism of Mendelian Inheritance*, 1st ed., 1915, 2nd ed., 1922, New York: Henry Holt.

Muller, H. J. (1914), 'The Bearing of the Selection Experiments of Castle and Phillips on the Variability of Genes', *American Naturalist* 48: 567—576.

Muller, H. J. (1916: I—IV), 'The Mechanism of Crossing Over I—IV', *American Naturalist* 50: I: 193—221, II: 284—305, III: 350—366, IV: 421—434.

Muller, H. J. (1920), 'Are the Factors of Heredity Arranged in a Line?', *American Naturalist* 54: 97—121.

Muller, H. J. (1950), 'Our Load of Mutations', *American Journal of Human Genetics* 2: 111—176.

Owen, A. R. G. (1950), 'The Theory of Genetical Recombination', *Advances in Genetics Research* 3: 117—157.

Plough, H. H. (1917), 'The Effects of Temperature on Crossing Over', *Journal of Experimental Zoology* 24: 148—193 (data appendix: 194—209).

Provine, W. B. (1986), *Sewall Wright and Evolutionary Biology*, Chicago, University of Chicago Press.

Sarkar, Sahotra (1989), Reductionism and Molecular Biology, Ph. D. dissertation, Department of Philosophy, The University of Chicago.

Sarkar, Sahotra (1992), 'Models of Reduction and Categories of Reductionism', *Synthese* 91: 167—194.

Schank, Jeffrey C. (1991), Computer Simulation and Experimental Design in Biology, Ph.D. dissertation, Committee on the Conceptual Foundations of Science, The University of Chicago.

Sinnott, E. W. and L. C. Dunn (1932), *Principles of Genetics*, New York: McGraw-Hill.

Snyder, Lawrence H. (1935), *The Principles of Heredity*, Boston: D. C. Heath.

Stern, Curt and Anna Sherwood (1967), *A Mendel Sourcebook*, New York: Schocken Books.

Strickberger, M. (1968), *Genetics*, New York: MacMillan.

Sturtevant, A. H. (1913), 'The Linear Arrangement of Six Sex-Linked Factors in *Drosophila*, as Shown by Their Mode of Association', *Journal of Experimental Zoology* 14: 43—59.

Sturtevant, A. H. (1914), 'The Reduplication Hypothesis as Applied to *Drosophila*', *American Naturalist* 48: 535—549.

Sturtevant, A. H. (1919), 'Inherited Linkage Variations in the Second Chromosome', Carnegie Institute of Washington Publication # 278, Part III, pp. 305—341.

Sturtevant, A. H. (1921b), 'A Case of Rearrangement of Genes in Drosophila', Proceedings of the National Academy of Science 7: 235—237.

Sturtevant, A. H. (1926), 'A Crossover Reducer in Drosophila melanogaster Due to Inversion of a Section of the Third Chromosome', Biologische Zentralblatt 46: 697—702.

Sturtevant, A. H. (1931), 'Known and Probable Inverted Sections of the Autosomes of Drosophila melanogaster', Carnegie Institute of Washington Publication # 421, pp. 1—27.

Sturtevant, A. H. and G. W. Beadle (1939), An Introduction of Genetics, Philadelphia: W. B. Saunders. (facsimile reprint by Dover Books, 1962.)

Sutton, W. S. (1903), 'The Chromosomes in Heredity', Biological Bulletin of the Marine Biological Laboratory at Woods Hole 4: 231—248.

Trow, A. H. (1913), 'Forms of Reduplication — Primary and Secondary', Journal of Genetics, 2: 313—324.

van Fraasen, Bas (1980), The Scientific Image, Oxford: The Clarendon Press.

Waters, K. (1990), 'Why the Anti-Reductionist Concensus Won't Survive: The Case of Classical Mendelian Genetics', in A. Fine, M. Forbes, and L. Wessels, eds., PSA-1990, vol. 1, East Lansing, Michigan: The Philosóphy of Science Association, pp. 125—139.

Weinstein, A. (1918), 'Coincidence of Crossing Over in Drosophila melanogaster (Ampelophelia)', Genetics 3: 135—159; tables to p. 181.

Weinstein, A. (1935), 'The Theory of Multiple Strand Crossing Over', Genetics 21: 155—199.

Wimsatt, W. C. (1976a), 'Reductionism, Levels of Organization and the Mind-Body Problem', in G. Globus, G. Maxwell, and I. Savodnik, eds., Consciousness and the Brain, New York: Plenum, pp. 199—267.

Wimsatt, W. C. (1976b), 'Reductive Explanation: A Functional Account', in A. C. Michalos, C. A. Hooker, G. Pearce, and R. S. Cohen, eds., PSA-1976 (Boston Studies in the Philosophy of Science, volume 30) Dordrecht: Reidel, pp. 671—710.

Wimsatt, W. C. (1981a), 'Robustness, Reliability and Overdetermination', in M. Brewer and B. Collins, eds., Scientific Inquiry and the Social Sciences, San Francisco: Jossey-Bass, pp. 124—163.

Wimsatt, W. C. (1987), 'False Models as Means to Truer Theories', in M. Nitecki and A. Hoffman, eds., Neutral Models in Biology, London: Oxford University Press, pp. 23—55.

JAMES F. CROW

SEWALL WRIGHT'S PLACE IN
TWENTIETH-CENTURY BIOLOGY*[1]

Sewall Wright died on March 3, 1988, at the age of ninety-eight.[2] He had lived through the whole history of Mendelian genetics. He had outlived by more than twenty years the other members, R. A. Fisher and J. B. S. Haldane, of the great triumvirate who ruled theoretical population genetics for decades. Failing eyesight diminished his productivity in the last years of his life, but not his deep interest in evolutionary problems and in his shifting balance theory. His last article was on this subject and appeared in the January 1988 issue of *The American Naturalist*.[3]

Graduate students enjoy judging their elders; at least, my generation did. A frequent topic of leisure-time discussion was the greatest living American geneticist. Who might it be? Such names as L. J. Stadler and A. H. Sturtevant always came up, but the field usually narrowed to two — H. J. Muller and Sewall Wright.[4] Muller discovered radiation mutagenesis and contributed greatly to the technology of *Drosophila* research. He had a remarkable capacity to keep the entire field in mind; any observation or experiment led him to connections and implications far beyond the immediate ones. In this way, he stimulated and guided the thinking of a generation of American geneticists. Wright added another dimension: he brought a mathematical approach to the study of gene action, inbreeding, quantitative traits, animal breeding, natural populations, and evolution. He was also the preeminent mammalian geneticist of his time. Together, Muller and Wright provided intellectual leadership for the whole field of genetics. Their counterpart in Britain was J. B. S. Haldane, a polymath who played a guiding and synthesizing role but, unlike Muller and Wright, made no important experimental contributions.

After World War II, the role of the intellectual leadership in basic genetics was taken over by Joshua Lederberg and Max Delbrück who, together, forced attention on the nature of the gene itself — something that was at last coming within the geneticist's reach. Now, following the Watson-Crick model of DNA structure and the explosion of molecular biology, the leadership is not so clear; it has become much more

Sahotra Sarkar (ed.), The Founders of Evolutionary Genetics, 167—200.
© 1992 *Kluwer Academic Publishers. Printed in the Netherlands.*

diffuse. The field of genetics has advanced more by new techniques and experimental finesse than by the imaginative view of the broad issues and central questions that the earlier leaders displayed.

Wright is best known to biologists for his emphasis on population structure in evolution, embodied in his shifting balance theory. But his interests were much broader. He developed statistical techniques, especially path analysis; he was the principal contributor to the early development of mammalian genetics; he, along with Fisher, laid the foundations for scientific animal breeding; he quantified the study of inbreeding and population structure — his coefficient of inbreeding is a standard part of elementary transmission genetics; he was one of a small number of biologists who wrote technical articles in philosophy; and, of course, he contributed richly to the mathematical theory of population biology and microevolution, arguably the most successful mathematical theory in biology.

Wright was always shy and retiring. Although he published extensively and on occasion strongly defended his views against criticisms, he played no major role in popularizing his ideas. This was left to others. In two important areas he had strong and influential protagonists. In animal breeding, Jay L. Lush dominated the field for many years and his students spread throughout the world, carrying the Wright gospel with them. In evolution, Theodosius Dobzhansky played Huxley to Wright's Darwin; a prolific writer, clear expositor, and popular speaker, Dobzhansky did more than anyone else to bring Wright's views to evolution-minded biologists. Wright's work would have been recognized without these two, but surely not so widely nor so quickly.

Wright was born in Melrose, Massachusetts, and grew up in Galesburg, Illinois, where his father taught at Lombard College. After graduating from Lombard, Wright got a master's degree at the University of Illinois, and then received his Sc.D. from Harvard working with William E. Castle. After leaving Harvard, Wright held only three positions: senior animal husbandman at the United States Department of Agriculture (1915—1925), professor at the University of Chicago (1926—1954), and professor at the University of Wisconsin (1955—1960). He continued active research and writing as professor emeritus until shortly before his death. This spindly skeleton is fully fleshed out in William B. Provine's book.[5]

WRIGHT'S SERVICE TO OTHERS

It may seem strange to those who knew Wright only through his extensive publications to start this article by emphasizing his service side. Yet, this was a major part of his life, and a major drain on his time and energy. During his first ten years at the University of Chicago a typical year's teaching consisted of two courses in each of two quarters and one in another, leaving only one quarter free; in three of the Chicago years he taught all four quarters. He also had graduate students and postdoctoral fellows. Wright was an exceedingly conscientious teacher: he prepared his lectures carefully, and they were detailed, comprehensive, and up to date.[6] His courses were often attended by faculty and visitors — Leo Szilard and James Watson, to mention two. Wright taught his own laboratories. His famous 1931 paper was delayed several years because of heavy teaching responsibilities during his early years in Chicago.

Another way in which Wright was of great service was as a manuscript referee. Several journals used him as a reviewer. When he received a manuscript he tried to make the review his first priority, usually putting aside his own research and writing. Always he did a conscientious job; and a conscientious job on a mathematical or statistical manuscript meant redoing the calculations and verifying the formulae. Wright's style was to derive the formulae himself, rather than to follow the method of the original author. A Wright review was often several pages long and extremely thorough; moreover, it was always constructive. I recall one instance in which Wright, as an anonymous reviewer, spent an enormous amount of time reanalyzing the data of a large experiment in quantitative genetics of the mouse and reached conclusions opposite to those of the original author. The author simply rewrote the paper with a new conclusion.[7] Wright must have done this sort of thing countless times. The editors of *Genetics* tried to avoid overworking him with manuscripts; yet, he still got the tough ones, especially those that were mathematical. It was automatically assumed by most readers and writers that any mathematical manuscript had been reviewed by Wright.[8]

Many people sent Wright drafts of manuscripts or data for analysis. This was by far the most time-consuming of his service activities, for he always did a thorough, thoughtful job, often involving extensive calculations. Every inquiry, every request brought a detailed response. He did a number of analyses of Carl Epling's data on *Linanthus*, and his collaboration with Dobzhansky is well known.[9] In these and other instances Wright spent inordinate amounts of time in analysis and correspondence. Inhabitants of the Genetics Building at the University of Wisconsin would regularly see him working, hour after hour, at his Monroe calculator. Often he entered these collaborations reluctantly, since they took time from his own researches; but he was extremely conscientious, and found it very hard to say no. Provine has detailed the difficulties Wright had in terminating his entangling alliance with Dobzhansky.[10]

When Wright was employed as senior animal husbandman by the U.S. Department of Agriculture, part of his job was answering inquiries. Many of these came from practicing livestock breeders and others not scientifically trained. The answers were given conscientiously and fully with the kind of thoughtful knowledge that only Wright possessed. His publication plans were often several years behind because of the presure of such duties.

Despite these distractions, Wright's total output was enormous — 211 scientific publications, almost all authored by him alone. He published four books, all after age seventy-five. Only after two retirements — from Chicago at age sixty-five, and from Wisconsin at seventy — could he find time for this, his major writing project.[11] One can only wonder what he would have produced had he followed the all-too-common practice of neglecting teaching and service to others and selfishly pursuing his own research interests. The time spent reviewing and improving often-mediocre manuscripts might have been better spent on his own research.

STATISTICS

Wright was always interested in mathematics.[12] Yet he had very little formal training in it — nothing beyond elementary calculus, learned from his father at Lombard College; the rest was self-taught. Between his junior and senior years in college he spent a year on a railroad surveying crew in South Dakota, where his mathematical skills were useful for determining the spiral curves involved in laying the rails around bends. During this time he contracted pleurisy and, confined to a railroad car, used the time to study quaternions.[13] He learned statistics by reading the literature, particularly the work of Karl Pearson.

As early as 1917, in an analysis of guinea pig weights, Wright discovered how to subdivide product moments into within- and between-strain components; he had independently discovered the analysis of covariance.[14] In 1920, while analyzing white-spotting in guinea pigs, he found a transformation to linearize cumulative percentage data; he had independently discovered the probit transformation.[15]

Wright's greatest contribution to statistics is his method of path analysis.[16] Although the mathematics are those of partial regression, the point of view is highly original and characteristically Wrightian. Wright was always interested in using statistics interpretively rather than simply descriptively, and he regarded significance tests as necessary but not very interesting. He was dissatisfied with correlation and regression coefficients as descriptive or predictive statistics; he was interested in causal analysis. His method was useful in situations in which the causal pathways were known, or could be plausibly assumed, but their relative importance was unknown. He liked to diagram causal sequences, using arrows to indicate causal relations and double-headed, curved arrows for correlations whose causal relationship to other variables was obscure or need not be considered. This was a natural way to diagram genetic relations in pedigrees, but Wright soon extended it to other problems. Each step in a causal pathway was associated with a "path coefficient," a partial regression coefficient standardized by being measured in standard deviation units; the path coefficients were then measures of the relative contributions of individual steps in causal paths. Wright was able to devise simple rules by which it is easy to write the relevant equations, which can be solved for the unknown variables. The method has the virtue of immediately making obvious whether there are sufficient measurements and relationships to permit a unique solution.

The most impressive of the early analyses using this method was Wright's study of production and prices of corn and hogs, done in the early 1920s. There were 510 correlations to be analyzed, and Wright did the calculations himself — a job that would be easy now, but was difficult and enormously time-consuming with the primitive card-sorting and calculating equipment then available. He was able to account for 80 percent of the variance of hog production and prices in the period between the Civil War and World War I by fluctuations in the corn crop and the various intercorrelations. A particular innovation was the introduction of time lags. Modern researchers are appalled to learn that the Department of Agriculture did not permit Wright to publish the paper; it was not proper for an animal husbandman to be meddling in economic problems. The paper was not published until a change in the national administration. Henry Wallace, himself interested in correlations, was impressed by the analysis and persuaded his father, then-secretary of agriculture, to intervene and get Wright's paper published.[17]

Among other applications of the method that Wright used were: analysis of inbreeding and assortative mating in livestock, correlations of prey and predator, respiratory physiology in humans, transpiration in plants, population structure and migration, isolation by distance, and the relative importance of heredity and environment in determining IQ. The method forms the basis for Wright's classic papers on livestock breeding and population structure.

A striking aspect of path analysis is that, aside from animal breeders, almost no one made use of the method for more than forty years. The important uses and innovations were accomplished mainly by Wright himself. Why? One reason is that the method cannot be used blindly and routinely, the way many standard statistical procedures are; it doesn't lend itself to "canned" programs. The user has to have a hypothesis and must devise an appropriate diagram of multiple causal sequences. Another reason is that most biologists have not been quantitative, while quantitative research in physics and chemistry was usually designed according to other paradigms. The explanation does not lie in the failure to use correlation and variance analysis, for these have been very popular, but the interest was usually in prediction equations, estimation procedures, and significance tests. Most researchers have preferred the neat layout and mechanical calculations of Fisher's analysis of variance and factorial designs. At the same time, psychologists have preferred factor analysis, a method that is mathematically similar to path analysis, but conceptually quite different.

One defect of path analysis, until recently, has been the absence of methods for assessing the statistical significance of the coefficients. Another, perhaps deeper, reason is that Wright's procedures were intuitive and heuristic. He didn't state his assumptions in a mathematical way, nor attempt a rigorous justification. Statistically minded biologists prefer mathematical to diagrammatic models, and often were not willing to accept Wright's methods without proof. But for many nonmathematical biologists, the problem was the opposite — the approach, for them, was *too* mathematical.

Yet, I think that the main reason why the method was not more widely used is that there was no program of advocacy. Wright wrote no textbook or instruction manual. Only animal breeders, because of Lush, had any systematic training in the method. Population geneticists not in the Lush tradition found the first textbook account in a book by C. C. Li, published in China in 1948.[18]

The path analysis method has recently become popular in the social sciences. The number of papers in education, econometrics, political science, sociology, and psychology using this method increased enormously in the 1970s, and path analysis is now de rigueur in some fields. The popularity arises not only from its obvious adaptability to non-experimental data such as are often found in the social sciences, but also from the method's being championed by such influential figures as Otis Dudley Duncan, Tad Blalock, and Arthur Goldberger. Wright's name may well become as widely known in the social sciences as in the biological.[19]

New methods and machine.computation have made Wright's methods more powerful. Yet he was not satisfied that all the applications adhered to the high standards that he himself set. At the same time there have been criticisms of the method, and one of Wright's last papers was a spirited defense of path analysis against some mathematical criticisms.[20]

ANIMAL BREEDING

While a graduate student at Harvard, Wright worked with guinea pigs, especially coat color patterns. On moving to the U.S. Department of Agriculture in 1915 he took charge of the ongoing guinea pig breeding program. The colony included 23 lines that had been sib-mated for many generations. Wright inherited not only the animals, but a set of well-kept records. These included, in addition to pedigree information, an abundance of measurements on such things as weight at various ages, litter size, and viability — the kind of traits that animal breeders were interested in — as well as color pattern. Wright was able to extract an enormous amount of reliable and consistent information despite unsatisfactory husbandry conditions (the animals had to survive poor quarters and wartime shortages, and especially the Washington summer heat). His statistical expertise is apparent throughout the analysis, for example in his judicious seasonal adjustments.

Wright's analysis of the effects of inbreeding and crossbreeding, published in 1922, reads as if it were written today.[21] He understood essentially all that we now know about the effects of inbreeding: random fixation of different traits in different lines; usual but not invariable decline in vigor and other fitness traits;[22] the recovery of vigor on crossing inbred lines; and the *quantitative* predictability of decline when these crossbreds were inbred. He attributed the entire effect, correctly, to Mendelism and dominance. He was insightful enough to realize that the degree of dominance, including overdominance, did not alter the linearity of inbreeding decline; epistasis was required to explain any nonlinear effects. In 1920 Wright wrote a bulletin for animal breeders. Here he suggested, as a breeding practice to gain control over weakly heritable traits, developing inbred lines to fix desirable traits and crossing them to get hardiness and vigor.[23]

Wright's most widely used result is his algorithm for computing the inbreeding coefficient, first presented in 1922 and now a routine part of elementary genetics courses.[24] It made possible the easy calculation of the decrease in heterozygosity for any pedigree, however complex, and has since become a standard part of the breeder's bag of tricks. Wright used this theory to analyze the history of Shorthorn cattle, a study in which he introduced a new wrinkle — namely, the division of inbreeding into two components, that due to consanguineous matings and that due simply to the small size of the breeding population. His analysis showed that the latter was by far the more important. He also wrote about the inheritance of quantitative traits. The symbol h^2, universally used to designate heritability, comes directly from Wright; it is the square of the coefficient associated with the causal path from genotype to phenotype, and is thus a measure of the relative contribution of genotypic differences to the variance of the trait.

As I said earlier, Wright — and the animal breeding community — profited greatly from the influence of Jay L. Lush. Lush quickly perceived the importance of Wright's work, spent a period of time with him, and soon became the recognized leader in animal breeding. He wrote the standard textbook in the field.[25] His influence, both directly and through his students, was worldwide. As a result, Wright's methods, particularly the quantitative theory of inbreeding and the use of path analysis, spread through the animal breeding community, and livestock breeding throughout the world changed from an art to a quantitative science.

In recent years, especially with the coming of artificial insemination in dairy cattle, the statistical methods applied to animal breeding have become much more sophisticated and computer-dependent. Current procedures for optimizing the effectiveness of selection, such as the BLUP (best linear unbiased predictor) method, bear little superficial resemblance to path analysis[26] — yet they are the direct descendants of Wright's pioneering methods. According to Provine there is no group in which Wright and his work are as revered as among animal breeders. Recently there has been a resurgence of interest in this method by human geneticists and students of morphological evolution.

PHYSIOLOGICAL GENETICS

Most biologists know of Wright because of his contributions to animal breeding and evolution. Yet a major emphasis of his work, *the* major emphasis for many years, was physiological genetics. The time spent studying his guinea pigs — he did the husbandry and record-keeping himself — and analyzing his data must have occupied the largest single time-block of his Washington and Chicago life.

Wright did not really choose to study guinea pigs; he was assigned to work on them by his major professor at Harvard, William E. Castle. Castle's laboratory included mice, rabbits, and guinea pigs, and Castle had a student for each. C. C. Little worked on mice, Castle himself (with E. C. MacDowell) studied rabbits, and John Detlefson, guinea pigs. Wright arrived as Detlefson was finishing and took over his work.[27]

Soon after moving to the U.S. Department of Agriculture, Wright wrote a series of papers on comparative coat-color inheritance in mammals. Included were individual articles on mice, rats, guinea pigs, rabbits, cattle, horses, swine, dogs, cats, and humans. These analyses were remarkable in two ways: they showed the striking similarities and presumed homologies of genes in the various species, and they were ahead of their time in interpreting the results in terms of the latest knowledge of enzymes and pigment chemistry.[28]

Wright's quantitative proclivities appeared early, in a paper analyzing the inheritance of size factors and the correlation of various body parts.[29] He was able to analyze the variance into components of general size, limb-specific factors, fore- and hind-limb factors, upper- and lower-limb factors, and factors specific to each part. The paper was the forerunner of many similar studies by others and, as usual, was ahead of its time. Such problems are now attracting a great deal of attention from developmental geneticists.

Gene interaction was a popular subject in the years after World War I. One of the most influential papers was Muller's, written in 1932, in which he used duplications and deletions to deepen the understanding of gene action in terms of the normal allele.[30] Normal genetic breeding experiments can only substitute one gene for another; they can't add or subtract. Muller showed how to get around this limitation by using small deficiencies and duplications. Muller's terms are still part of the genetics vocabulary: for example, a *hypomorph* is a mutant whose phenotypic action is in the direction of the normal allele, but less effective; an *antimorph* acts in a manner opposite to that of the normal allele. Wright carried this classification further. He was characteristically thorough and inventive in his analysis of guinea pig coat colors, and he wanted to explain dominance and epistasis in chemical terms. Those who know Wright's quantitative bent will not be surprised that he formulated the relations in path diagrams and expressed the processes as differential equations; what is, to me, more surprising is his skillful use of then-current knowledge of enzymes. The result was a masterful analysis of coat color interactions in terms of enzyme kinetics. He also deepened Muller's concept by showing that two dimensions were needed, corresponding to substrate-combining and turnover efficiency. From this he predicted another type, what he called a *mixomorph*: a gene producing an enzyme product with high combining but low turnover capacity. Such a mutant would behave as a hypomorph with mutants lower than itself on the scale of activity, but as an antimorph with higher alleles. An example was later found in *Drosophila*.

Wright's major analyses of this type were published in 1941.[31] That was the year that genetics took a new turn, when George Beadle and Edward Tatum's discovery of biochemical mutants in *Neurospora* opened up a new and more effective way of studying the relation between genes and enzymes. It can be regarded as the starting point of molecular genetics, for the *Neurospora* methods led directly to Joshua Lederberg and Tatum's discovery of sexual reproduction in bacteria. Wright continued to do guinea pig studies until his move to Wisconsin in 1955, but the analyses, inventive, careful, and detailed as they were, attracted little attention. The times had changed and the emphasis was on microorganisms, molecular biology, and the nature of the gene itself, not gene interaction based on the quantitative analysis of phenotypes.

POPULATION GENETICS

Any mention of population genetics automatically calls to mind three names: Haldane, Fisher, and Wright; for a generation they completely dominated the field. Each made unique contributions, and together they developed a mathematical theory that ranks with those of the physical sciences and is, I believe, unique in biology. Each of the three had a highly individual style; their papers would be instantly recognizable if the names were removed.

They each had other interests. Wright's I have mentioned. Fisher was the outstanding statistician of his time, perhaps of all time. Haldane is harder to describe. He was as close to the ideal of a polymath as anyone in modern science. Trained in classical languages, he was an author of science fiction, he was a left-wing political activist, and he was a newspaper columnist. He wrote popular science with remarkable clarity. The topics of technical papers included cosmology, enzyme kinetics, respiration, mathematics, statistics, demography, ecology — and of course, genetics.

All three men showed precocious mathematical ability. Fisher majored in mathematics at Cambridge and had additional training in mathematical physics. Haldane also had considerable mathematical training. Wright, as we have seen, was mainly self-taught, and yet he was undaunted; somehow he was able to get satisfactory, if not the best, answers to important questions, often by unorthodox methods. As to the kinds of problems they solved, Haldane's were relatively pedestrian; Wright solved more difficult problems and showed greater originality. Fisher's mathematical work in both statistics and genetics was far more powerful and original than either of the other two, and the problems that he solved were more difficult; time and again he found new and inventive approaches to seemingly impossible problems.

Haldane's early contributions included a series of papers that systematically explored selection under a wide variety of circumstances. He summarized his findings in 1932 in a book that is remarkable, not only for its insights, but for its lucidity;[32] it is at once good science and clear exposition at their best. Haldane is best known for two papers. In one he showed that the impact of recurrent mutation on fitness is a function of the mutation rate, and not of the degree of deleteriousness of the mutations themselves; in the other, he showed that the total amount of reproductive excess required to carry a new favorable mutant to fixation, while maintaining the population size, is a function of its initial frequency (and dominance).[33] Although he sometimes dealt with polygenic models, his main emphasis was on single loci, and interactions of a small number.

Fisher's mathematics differed stylistically from Haldane's and Wright's, being much more elegant. Haldane's characteristic style was to set up a problem, grind out the results as fast as possible, and publish them with no further polishing. He worked out the consequences of selection for a large number of modes of inheritance. Wright concentrated more on the central issue, as he saw it. His object was to find an answer; he did not much care how he got it, and often the solution was cumbersome. In contrast, Fisher's style was to invent a tricky, novel, elegant, and often ingenious approach. I believe that aesthetics were important to him; this shows in both his mathematics and his prose. Elegance took precedence over clarity, with the consequence that much of his writing is almost impossible to understand. Fisher introduced a stochastic element into the theory, showing for the first time the effect of population size on gene frequency drift (although he made a twofold error, later corrected by Wright); yet he analyzed this effect, only to dismiss it. In Wright's view, Fisher threw away the wheat with the chaff.

Fisher's best-known and most important contribution to evolution was his showing how to deal with gene interactions in quantitative traits, especially when the trait is fitness itself.[34] He regarded dominance and epistasis as ever-present nuisances. As early as 1918 he measured their effects on the correlation between relatives.[35] He showed that natural selection can pick out the additive component of the genetic variance, and that the rate of progress under selection is proportional to this component. He stated this in his 1930 book in the form of a 'Fundamental Theorem of Natural Selection': the rate of change in fitness is equal to the additive variance at that time. A few pages later Fisher introduced additional terms that make the theorem more realistic, but it is usually stated in its simplest form — which has led to some confusion in the literature. As usual, Fisher's explanation was not easy to understand, and a later derivation by Motoo Kimura is more direct.[36]

To Fisher, gene interactions and random gene frequency fluctuations were impediments to the progress of selection, and he regarded them much as one would noise in a physical system. To Wright, on the other hand, these provided opportunities for evolutionary creativity. I shall defer till the next section a discussion of this aspect, which is central to Wright's thinking.

From his interest in inbreeding and population structure, Wright developed his widely used F statistics. These are an extension of his inbreeding coefficient to include hierarchical population structure. First developed in connection with his studies of the history of Shorthorn cattle, they now form the basic framework for descriptive statistics of geographically structured populations.[37]

Wright's aim in his stochastic theory was to bring into one expression the influences of mutation, migration, selection, and random processes. His method was to look for an equilibrium solution. It is interesting to read Wright's papers chronologically, as he successively discovered more general ways of treating the problem, culminating in the Kolmogorov forward equation.[38] With each successive improvement he was able to take more factors into account, such as various levels of dominance and fluctuating selection coefficients. This theory represents the high point of Wright's mathematical work.

In a general, one-locus, two-allele case the Wright equilibrium equation is

$$\varphi(q) = C\overline{W}^{2N}q^{4N(m\bar{q}+v)-1}(1-q)^{4N[m(1-\bar{q})+u]-1},$$

in which q is the allele frequency in the subpopulation; $\overline{W}$ is the mean fitness, expressed as a function of the genotypic fitnesses and allele frequency; N is the effective subpopulation number; m is the proportion of the subpopulation that is replaced in each generation by migrants from the entire population whose frequency is $\bar{q}$; u and v are forward and reverse mutation rates; and C is a normalizing constant to make the sum over all values of q equal to one. Then $\varphi(q)$ is the probability that the subpopulation frequency is q.[39] This equation is basic to Wright's discussion of his shifting balance theory of evolution, to which I shall return in the next section.

The field of mathematical population genetics has advanced a great deal beyond the work of the three founders. The first to develop stochastic concepts in a rigorous way was Gustave Malécot. He gave the work of Wright and Fisher a much more solid justification, but recognition has come slowly because he published in French and in journals not familiar to most geneticists.[40] Starting in the mid-1950s the field began to be dominated by Motoo Kimura, who was especially adept in applying the partial differential equations of diffusion theory to population genetics. In particular, although Wright and Fisher had dealt mainly with equilibrium distributions, Kimura was able to construct the whole process from an arbitrary starting point, and to include Wright's results as special cases. More recently, other mathematicians — notably Samuel Karlin and Thomas Nagylaki — have entered the field, and it has become a very sophisticated branch of applied mathematics, no longer accessible to most evolutionary biologists.[41]

Although Wright's major contributions to population genetics were theoretical, he also participated in some experimental research. Studies of natural populations had traditionally been largely descriptive and phenomenological. Wright's first paper with Dobzhansky was the beginning of a new trend toward theory-guided experimental studies of natural populations.[42]

Wright's role as one of the three founders of population genetics is only part of his work. During his retirement years he wrote a four-volume treatise that not only summarizes his own work and adds some new results, but summarizes the work of Haldane and Fisher, as well as more recent contributions to the theory.[43] It is also loaded with experimental results, most of them reanalyzed and reinterpreted by Wright.

One cannot discuss Wright's role in population genetics and evolutionary theory without mentioning the name of Theodosius Dobzhansky. Dobzhansky was for many years the leading figure in experimental population genetics. Although he could not understand Wright's mathematical work, he was enormously impressed by it; during the 1930s, when the biological world was beginning to be interested in putting genetics and evolution together, he brought Wright's name to the general biological public. Dobzhansky never ceased to praise Wright and to seek collaborations with him. He was a great popularizer, and his *Genetics and the Origin of Species* brought Wright's work to the attention of a whole generation of evolutionists.[44]

THE SHIFTING BALANCE THEORY

Wright is most widely known for his view of evolution, which in later writings he called the 'Shifting Balance Theory.' It was an attempt to bypass some of the problems encountered by mass selection in a panmictic population by introducing a balance of random gene frequency drift, migration, and selection within and between groups. Wright formulated the idea before 1930 and adhered to it in almost its original form throughout his life.

Origins of the Theory

The shifting balance theory traces back to Wright's thinking while he was employed by the U.S. Department of Agriculture. It would be easy for one familiar with the theory to trace its origin, even if Wright had not himself done so, which he did.[45]

Four interrelated observations provide the background:

(1) The first came from Wright's experience as a graduate student at Harvard, where he assisted William Castle in his classical experiments selecting for increased and decreased amounts of white in hooded rats. In these experiments Wright saw that a few generations of selection had produced strains that were almost completely dark or white; at the same time, he was impressed by the eventual necessity to stop the experiment because of infertility. Intense selection for one trait had brought along other, unwanted ones.

(2) The second observation grew out of Wright's extensive analysis of hair color and pattern in guinea pigs. These studies started with his Sc.D. thesis and continued until his retirement from the University of Chicago, some forty years later. Wright was impressed by the frequently unpredictable effects of genes in combination. In one rosette pattern, for example, the combined effect of two genes was opposite to what would be expected from their individual effects.

(3) The third observation came from the studies of inbreeding and crossbreeding in guinea pigs. It was already well known that inbreeding usually leads to a decline in vigor and fitness, but Wright was also struck by the uniqueness of each line — not only in superficial aspects, such as color pattern, but in internal organs, physiology, behavior, and fertility. Inbreeding had enhanced random differentiation among the lines and fixed different combinations in each one. The variability

among inbred lines was much greater than that in the foundation stock from which the lines were derived.

(4) The fourth observation came from Wright's studies of the history of domestic livestock, especially Shorthorn cattle, for which the records went back many years. Here he observed that the general advance in the quality of the breed appeared not to have come from systematic selection by breeders. Rather, he found that occasionally, and for no obvious reason, a particular herd showed superior qualities. Bulls exported from such a herd transmitted their genes to the whole breed, so that in a few generations it attained the quality of the best herd. Then the cycle was repeated.

The selection of hooded rats showed Wright that mass selection could be effective in changing a particular trait, but that intense selection brought along unwanted traits as passengers. The guinea pig coat colors showed him that selection for individual traits might not produce the best combinations. The inbreeding experiments showed that mostly unfavorable combinations, but occasionally favorable ones, were fixed in the line; and that inbreeding within lines increased the variance of the whole set of lines. Finally, the Shorthorn cattle history showed that by repeatedly bringing in sires from a particularly good herd, the breeder could bring his herd up to the quality of the good one.

Replace "guinea pigs" and "cattle" in these statements with "natural populations" and you have the essentials of Wright's shifting balance theory. He arrived at this view in the early 1920s, but his move to Chicago in 1925, with its new responsibilities, kept him from completing the analysis and publishing it. The first presentation, largely written by 1925 but not published until 1931, was his most famous paper: 'Evolution in Mendelian Populations.'[46] Although Wright refined and clarified his theory and the mathematics were generalized and made more rigorous, the theory itself hardly changed from its inception.[47]

What Is the Shifting Balance Theory?

The shifting balance theory was Wright's attempt to bring the rodent and livestock experience to bear on a major problem in evolution — the evolution of harmonious, coadapted gene complexes.

The problem. Wright consistently emphasized that often it may not be possible for a population to get from *here* to *there* by mass selection. A simple haploid model is sufficient to illustrate the problem. Suppose alleles *a* and *b* go well together, as do *A* and *B*, but *A* and *b* or *a* and *B* do not. Suppose further that the *AB* combination is better than the *ab*. If a population has a high frequency of *a* and *b* alleles it will not move to a state in which *A* and *B* are common, because to do so will produce a large number of inferior *Ab* and *aB* recombinants.

We can think of this as a three-dimensional graph in which the two abscissas are the frequencies of the *A* and *B* alleles and the ordinate is the mean fitness of a population with this frequency combination. The surface will be saddle-shaped, with a low peak where *ab* is common and a high one where *AB* is common. A population near the lower peak cannot get to the higher one without crossing a valley of lower fitness (Figure 1).

The mean population fitness could go from a position near the lower peak to the higher one if the genes were closely linked, but Wright regarded the probability that two or more favorably interacting alleles will also happen to be closely linked as too small to constitute a general explanation. But another way that the saddle could be crossed is for the population to be small. In such a population, as in inbred lines, selection is ineffective and allele frequencies drift more or less randomly. By

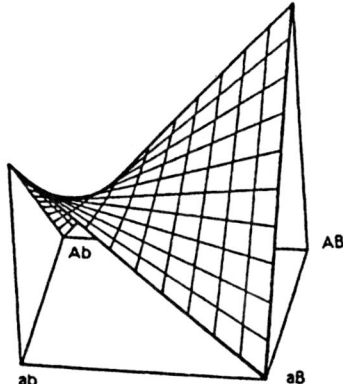

Fig. 1. A simple, haploid Wrightian adaptive surface. The abscissas are the frequencies of the *A* and *B* alleles; the ordinate is the mean fitness of a population with this combination of *A* and *B* allele frequencies. The two loci are unlinked.

chance, the small population frequencies might cross the valley to a higher peak.

The probability that this happy gene frequency shift will occur by chance is low. But if there are many such small populations, there is a reasonable chance that this might happen in one or a few of them. Wright viewed this situation as the most favorable for evolutionary creativity.

I have given only the simplest illustration. The fitness of a real population is determined by a large number of genes. The model can be generalized by thinking of a multidimensional abscissa, but with a single ordinate dimension still representing mean fitness of individuals in the population. The simple, saddle-shaped surface becomes a hypersurface with multiple peaks, ridges, and valleys. A peak is not necessarily at a corner, for in a diploid population the most fit genotypes at some loci may be heterozygous. An actual population, even a very large one, is not likely to be exactly at a peak because of recurrent mutation, migration, and environmental fluctuations.[48]

Despite the complexity, and the impossibility of drawing the appropriate multidimensional diagram, the dilemma is made clear by a simple extension of the simple example. A population with gene frequencies such that its average fitness is in the vicinity of a peak is quite unlikely to be near the highest peak. Selection in a large panmictic population cannot carry the population gene frequencies across a valley to a higher peak. Yet if the population were broken up into small populations, each subject to random changes, among these might be one (or more) that had drifted across a saddle into an area where selection could carry it to a higher peak.

Phase 1. The first phase of Wright's process is based on what I have just been describing. He visualized a large population, broken up into a large number of subpopulations, within which there are random changes.

If the subpopulations are completely isolated, each will tend to become homozygous, and very likely most of them will become extinct because of fixation of deleterious genes. For the Wright process to work there must be sufficient genetic variability within the subpopulations to permit gene frequency drift. Mutation rates are ordinarily too low to maintain variability in very small subpopulations. A more likely source of variability is migration from other subpopulations with different gene frequencies; furthermore, in contrast to new mutations, these genes are pretested, the most harmful ones having been eliminated by natural selection. Therefore the populations must not be completely isolated, only partially so. Yet, if migration is too great the whole population is essentially a single panmictic unit. So there is an optimum level of migration, sufficient to keep local populations from becoming fixed but small enough to permit considerable local differentiation.

Wright believed that most traits of evolutionary importance — size, blood pressure, hormone levels, stomach pH — are quantitative and multifactorial, and that an intermediate value is optimum. In such a system there are some alleles that increase the value of the trait and others that decrease it. This means that there are many gene combinations that produce essentially the same phenotype. In a geographically structured population each local subpopulation will tend to have its own constellation of gene frequencies, even though the average phenotypes will be rather similar; but since genes affecting a particular quantitative trait also have other effects, these subpopulations will differ somewhat in average fitness.

Selection for an intermediate optimum automatically introduces epistatic interaction of the genes involved. The fitness effect of a gene depends on what other genes are present; for example, a gene increasing blood pressure will be advantageous in an individual with low blood pressure, but detrimental in one with high pressure. That is to say, with an intermediate optimum, genes that are additive on the phenotypic scale are epistatic on the fitness scale. Thus, there is precisely the kind of interaction that makes it impossible for mass selection to carry the population from a low peak to a higher one.[49]

A large panmictic population will have various genotypes concentrated rather closely about one near-optimum genotype — which, however, need not be the most fit of all such near-optimum combinations. A Wrightian subdivided population generates a larger field of fitness variation among the subpopulations than would exist in a panmictic population (recall the high variance among inbred lines of guinea pigs), making it more likely that a better combination exists somewhere among them.

Wright emphasized that a subpopulation structure does not have to involve physical barriers or large gaps between subpopulations. A sparse population with limited individual mobility may have the equivalent of a substructure, leading to what he called "isolation by distance."

Although the mathematics become much more difficult, the qualitative conclusions are the same.

It is in the context of phase 1 that Wright did most of his theoretical work. He asked what effective subpopulation size (usually somewhat less than the number of breeding adults) and rates of migration, mutation, and selection permit an appropriate amount of trial-and-error gene frequency drift. If there is too much migration, all the subpopulations tend to be similar. If the subpopulations are too small, or the migration rate too low, they drift into homozygosity, and often, into extinction. The best chance for the trial-and-error process to produce a superior combination is when the various factors are in approximate balance.

The population structure must be such that the amount of migration per generation is on the order of the reciprocal of the local effective population number, or roughly a single migrant per generation. The selective differences among the genotypes must also be of the order of the reciprocal of the local population size, or less. The situation most favorable for a Wrightian balance is when the frequency function, $\varphi(q)$ (see above), is approximately flat.

Phase 2. Once the gene frequencies have drifted across a valley and into the domain of attraction of another peak, mass selection will carry them toward the peak. The random process doesn't have to carry the population to the higher peak, only across the valley; then selection takes over. This process, in which the favorable gene combination is increased by selection within a subgroup, is Wright's phase 2.

Phase 3. Assume that one of the subpopulations has drifted into a favorable combination of gene frequencies. This subpopulation, being more fit than the others, will increase in numbers relative to its neighbors. There may be competition between the subpopulations, with the weaker ones losing out; a much more likely result, however, is that, as the favored subpopulation grows, individuals from this group will emigrate and mate with members of surrounding populations. In a process fully analogous to "grading up" in animal breeding, the whole population will be brought up to the fitness of the favored subpopulation.[50]

Wright was very mathematical in his treatment of phase 1. Phase 2 requires no complex mathematics; it is simply mass selection. Surprisingly, Wright's treatment of phase 3 was entirely descriptive. What keeps recombination from breaking up the favorable gene combinations brought in by the migrants? The lack of quantification of phase 3 has been regarded as a major criticism of the theory. For example, Haldane said: "To my mind the weakest point in Wright's argument is that he has not adequately considered what happens when this [subpopulation] starts hybridizing with others."[51] Recent numerical analysis of realistic examples shows, however, that Wright's intuition was correct. Very small rates of migration, considerably less than the selective differences, are sufficient for the upgrading of the recipient populations, even with free recombination. The migration does not have to be unidirectional.[52] I conclude that the third phase is not the weak part of Wright's theory.

There is need for an extension of this analysis to more realistic situations, with multiple subpopulations having different gene frequencies, variable rates of migration, and various kinds of gene interactions; and a stochastic treatment is needed. But the main qualitative result seems clear already: migration at a realistic rate is an effective means of transforming the entire population to the genotypes of the best group.

Criticisms of the Theory

Criticisms of Wright's shifting balance idea usually fall into one of three general categories. The first is physiological. Wright's argument depends strongly on the observation of unpredictable gene combinations affecting guinea pig coats. This clearly suggests the "peaks and valleys" model. Yet multiple genes contributing to quantitative traits are usually lacking in dominance and show little epistasis — so one criticism is that Wright's mechanism may not be necessary. There may always be enough additive genetic variance for mass selection to be effective. Perhaps Wright was unduly influenced by the complicated interactions of guinea pig coat color genes. He emphasized, however (as mentioned earlier), that when an intermediate phenotype has the maximum fitness, alleles that are additive on the phenotypic scale are strongly interactive on the fitness scale. He also regarded pleiotropy as ubiquitous.

In terms of the peaks and valleys metaphor, it has been argued that, as the number of genes involved increases, the fitness hypersurface becomes one of low peaks, shallow valleys, and many more ridges than peaks. Furthermore, fluctuations in the environment may change individual fitnesses and make the fitness surface more like the undulating waves of an ocean than like a mountainous land surface. Thus it may be that a population never finds itself in such a position that *no* incremental change in allele frequencies can produce an increase in fitness. If so, then selection based on additive effects of genes would be the most effective way of increasing fitness and keeping up with an ever-deteriorating environment. This criticism came primarily from R. A. Fisher.[53]

H. J. Muller, although not opposed to Wright's model, especially as applied to the development of social behavior in early human evolution, was not so convinced of the restrictions imposed by pleiotropy. He argued that there are almost always alternative ways of accomplishing the same phenotypic effect. If there is a gene with two effects, then there will usually be modifiers that affect only one of them. I might paraphrase Muller's and Fisher's views this way: Muller would say that there will usually be other ways of going from one good phenotype to another that do not involve crossing a saddle. Fisher would say that the important question is not how to get to a specific phenotype, but how to increase fitness; although it may not be possible to go to a specified peak, it will in general be possible to go to something better than currently exists.

A second class of criticisms comes from the rather strict requirements of population structure for the shifting balance theory to work. One can question whether the proper balance of migration, subgroup size, and the other parameters required for phase 1 exists often enough to be important. A difficulty is our lack of knowledge of the history of innovative steps in evolution. If we knew enough, we might find that evolutionary changes combining several traits that appear to be individually disadvantageous have in fact involved numerous intermediate steps, each selectively advantageous.

A third class of criticisms is more Popperian. Wright always tried to encompass all relevant factors simultaneously. He preferred to present the theory in its full complexity, believing that any simplification loses the essence, which is the simultaneous interaction of many complex forces. This makes it difficult to test experimentally, for only by considering individual parts can one hope for specific quantitative predictions. Thus, Wright's shifting balance theory is not a mathematical theory with specific assumptions and predictions, but more like a metaphor. This, however, does not decrease its appeal to many biologists. It means, though, that evidence has to come from sources other than agreement with mathematical and experimental predictions.

Evidence for the Theory

Wright never tried to test his theory by a detailed examination of evolutionary patterns and population structures or by laboratory experiments, for the reasons just stated. He regarded the theory as too broad and general to be much affected by any single observation or experiment. As evidence, he relied on the rodent and cattle analogies and the inherent plausibility of the scheme.

Direct evidence is hard to obtain. The strongest evidence, in Wright's view, is that major evolutionary advances typically have left the poorest fossil records. This is consistent with the view that these took place when the population was sparse. Large populations that left abundant fossil records were changing in less fundamental ways, with fine-tuning rather than major innovations. Wright reiterated this argument in his last paper (1988).

Another source of evidence is the correlation between the rate of morphological change and the number of cytological changes. Mammals have evolved more phenotypic novelties and have changed structurally more rapidly than the lower vertebrates, and they have also had more chromosome rearrangements in their history. Inversions and translocations are deleterious when heterozygous, and it is hard to imagine an orderly selection process that would carry a population from having one gene sequence to having another; the intermediates would be at too strong a selective disadvantage. Cytological evolution seems to require a population-size bottleneck, or a population small enough to permit random drift despite strong selection. The observed correlation between cytological changes and rapid morphological and behavioral evolution (e.g., in mammals compared to lower vertebrates) offers support for Wright's idea.

The evolution of early humans and other primates is consistent with the Wright idea. The population structure was one of numerous very small colonies — a structure that fits the requirements of Wright's theory. And primate evolution, both morphological and behavioral, has been extremely rapid.

There has been interest in recent years in "punctuated equilibrium" — that is, periods of very rapid change interspersed between long periods of very little change. How abrupt such transitions are is a matter of dispute, but that rates of evolution vary enormously is not. Some "punctuationists" have invoked Goldschmidtian macromutations as the mechanism for rapid fundamental changes.[54] Wright preferred the view that these changes are polygenic and that the rapid changes, which have left no fossil intermediates, are the result of just such a sparse population as would produce rapid evolution on his theory.

The neutral theory should, I think, increase the acceptance of Wright's work. If there are many DNA changes that are so nearly neutral that they are fixed in evolutionary time by random drift *in the global population*, as Kimura argues convincingly,[55] there must be many genes weakly enough selected to be subject to random drift *in local subpopulations*, as Wright's theory requires.

Was Wright or Fisher Correct?

I should reemphasize that Wright and Fisher had different objectives. Fisher was interested in the ways in which populations increase in fitness — or, more accurately, change genetically fast enough to keep up with an ever-deteriorating environment. In contrast, Wright was always interested in the evolution of novelty, which may depend on specific interacting gene combinations. The sixty-year-old controversy is not resolved, and it is hard to imagine a definitive resolution coming soon.

In considering possible further evidence for Wright's mechanism there are two quite different kinds of empirical questions: (1) What kind of population structure is optimum for evolution? To answer this, one needs to know the extent to which complex interactions set limits to advance by mass selection, and we need more detailed population theory. (2) What are the actual structures of populations, and, more important, what were the structures in the past at the time that rapid evolutionary changes were occurring? To be useful, such information would require detailed data on a number of representative populations. The two questions — what would be best, and what actually exists — are both difficult to answer.

There have been recent attempts to extend mathematical theory to include adaptive surfaces and shifts among equilibria.[56] It is too early, for me at least, to judge how important these will be in the final assessment of Wright's theory.

I would like to think that the world is complex enough to encompass both ideas, and that they are more complementary than antagonistic. There seems to me to be little doubt that the fine-tuning of structures and functions for higher fitness, and the genetic changes required to track a changing environment, are well accomplished by mass selection acting on additive variance. So are long-term trends, such as increasing size. On the other hand, it may be that *real* novelties require unusual, interacting, lock-and-key mechanisms that cannot be built up by stepwise selection. For such as these, the Wrightian mechanism may be required. Furthermore, Kimura's work has made us realize that, quite aside from morphological and functional changes, there is a great deal of random evolution going on at the DNA level. All three processes are valid, although their relative importance remains to be determined — if, indeed, one can formulate the problem sufficiently precisely to permit a definitive answer.

PHILOSOPHY

In 1952 Wright was elected president of the American Society of Naturalists. His presidential address surprised everybody: he talked, not about evolutionary or physiological genetics, but on the philosophy of organism and on the mind-matter problem.[57]

Wright espoused an uncommon solution to the problem. He used the word "organism" to conform to Webster's broadest definition: "any thing, structure, or totality of correlative parts, in which the relationship of part to part involves a relationship of part to whole, thus making it self-inclusive and self-dependent."[58] An organism can then be an individual plant or animal, an ant colony, a human society, a species, an ecosystem, or the entire biota. Extending this to the physical world, it could be the earth, the solar system, or the universe. Going downward, Wright found molecules, electrons and protons, and (if he were writing today) quarks. He emphasized the continuity between individual, organ, cell, virus, and gene — and between living and nonliving.

With such a hierarchical organization and no clear borders between the different levels, Wright found it difficult to imagine truly emergent properties, such as mind. There is no place at which one can say that mind exists only beyond this point, whether one is speaking of time-points in phylogeny and ontogeny or of contemporary levels of organization, from electron through living organisms to the universe. Emergence of mind from no mind, he said, is "sheer magic": "If the human mind is not to appear by magic, it must be a development from the mind of the egg and back of this, apparently, of the DNA molecules of the egg and sperm nuclei that constitute its heredity" (p. 278). Seeing no stopping point here, Wright said: "Because of the hierarchic nature of biologic and physical entities it appears that my mind must be based somehow on the minds of my cells and these on those of their constituent molecules and so on down to elementary particles" (p. 283).

My mind sees yours as matter; yours sees mine as matter. My stream of consciousness is inaccessible to you, except as we compare experiences. Yet the power of mind is remarkable. Wright noted:

A human being appears to make choices that make enormous differences in the course of events. How can something that is composed entirely of molecules have a freedom that transcends the statistical regularity of an aggregation of molecules? The answer is, of course, that a tightly integrated organism operates through a hierarchy of switch or trigger mechanisms. In an action involving a very great exchange of energy, for example the flight of an aeroplane, all molecules may follow their customary modes of behavior, except in a small portion of the system (a lever and the pilot) which diverts the course of the whole in one way or another. Within this small portion there may in turn be no detectable swerving from customary behavior except in a switch mechanism of the second order of smallness (a neuro-muscular junction), and so on to transactions of the third order in neural and cerebral synapses. The portion of the total energy exchange involved in controlling the deviation from customary behavior may thus be of the nature of an infinitesimal raised to a high power. Yet the flight is all according to plan. (p. 286.)

Wright thus arrived at his view of "dual-aspect panpsychism" — that mind is everywhere. "The only satisfactory solution of these dilemmas would seem to be that mind is universal, present not only in all organisms and in their cells but in molecules, atoms and elementary particles" (p. 278). Mind is universal; so is matter.

Science, aside from what we can infer about minds from comparing notes about our perceptions or from introspection, deals with the external aspects. It gives us statistical descriptions rather than accounts of a deeper reality. He wrote:

Acceptance of this point of view requires relatively little change in the actual practice of science, especially as determinism has never been more than an ideal admittedly unrealizable in full because of the invariable errors of observation and in many cases, practically irreducible probabilities like those in the fall of dice (or segregation and assortment of genes). The deterministic expressions do not lose their usefulness as approximations. What we are given is a tenable philosophy of science and along with this a desirable humility in the recognition that science is a limited venture, concerned with the external and statistical aspect of events and incapable of dealing with the unique creative aspect of each individual event. (p. 288.)

The biological community has greeted Wright's philosophy sometimes with disagreement, but more often with indifference. Many biologists regard the mind-body problem as something to avoid, better left to philosophers. Others regard mind as something growing out of matter: the more complex the organization of matter, the more complicated the mind can be. Extending this view to the computer age, Marvin Minsky says: "This book assumes that any brain, machine, or other thing that has a mind must be composed of smaller things that cannot think at all."[59] Wright did, however, find kindred spirits among a few philosophers, especially his University of Chicago colleague, Charles Hartshorne. Of Wright's several articles developing these ideas, one was in a Hartshorne Festschrift.[60]

WRIGHT'S IMPACT

What have been Wright's major influences on twentieth-century biology? Here is one person's assessment:

Statistics. Path analysis continues to be a major method for analysis in poorly controlled situations where the causes are known or can be assumed, but their relative importance is not. The recent introduction of methods for assessing significance and computer routines for handling large bodies of data have increased its popularity. I expect an increasing use of path analysis in nonexperimental situations in both biological and social sciences. There has been a renewal of interest by human geneticists. The method has been criticized for insufficient attention to its logical foundations and basic assumptions; these can be clarified, and the method will then have a wider acceptance. As the situation becomes better understood, as has happened with livestock breeding, the formalism will change — but the roots are in Wright's path analysis.

Animal breeding. Unless molecular trickery becomes so powerful that breeding methods can be dispensed with — and this is not likely to happen soon, because most yield and performance traits are polygenic — Wright's methods and their derivatives will continue to dominate this field. Concepts of heritability, genetic correlations, selection indices, and prediction formulae grow directly out of his work. Wright's inbreeding and population structure theory is also becoming of increasing importance in devising strategies for the preservation of endangered species, both in nature and in zoos.

Physiological genetics. Although Wright's guinea pig papers are not often read or discussed currently, his work played an important part in bringing genetic analysis and enzyme chemistry together. The methods have been superseded by more powerful and direct approaches using microorganisms, cell cultures, and in-vitro chemistry. As more attention returns to the quantitative study of developmental genetics, Wright's methods may have a resurgence. An example of the fusion of Wright's methods with modern molecular knowledge is the work of H. Kacser and J. A. Burns.[61]

Population genetics. This is now a thriving science of its own, thanks to the pioneering work of Wright, Fisher, and Haldane. Wright's most important contributions were the quantification of population structure and migration and the emphasis on stochastic elements. His mathematical results are a part of the permanent heritage. Together with Haldane and Fisher he has laid a magnificent foundation for current and future work, both theoretical and experimental.[62]

Evolution. Wright's shifting balance theory, with its emphasis on complex gene interactions, random gene frequency drift, and geographically structured populations, has been, and continues to be, very attractive to biologists. Regardless of the judgment by future workers of its ultimate significance, it will have played a most important part in the theoretical developments of twentieth-century ideas of evolution. Wright, along with Fisher, has been responsible for the "modern synthesis," putting Mendelian inheritance and Darwinian natural selection together into a coherent body of knowledge.

In any one of these areas Wright ranks among the important contributors to twentieth-century biology. Considering them together, he must be ranked among the greatest. That his name is still very much alive is shown by the 1988 Science Citation Index, which lists some 500 articles that refer to his work.

ACKNOWLEDGEMENT

* Reprinted from the *Journal of the History of Biology* 23: 57—89 (1990). I am indebted to Thomas Nagylaki for clarification of ideas and several useful suggestions. This is contribution no. 2948 from the Laboratory of Genetics, University of Wisconsin.

University of Wisconsin

NOTES

[1] Two recent publications provide an abundance of information about Wright: William B. Provine, *Sewall Wright and Evolutionary Biology* (Chicago: University of Chicago Press, 1986); and Sewall Wright, *Evolution, Selected Papers*, ed. with introductory materials by William B. Provine (Chicago: University of Chicago Press, 1986). I have had more than thirty years of close association with Wright; nevertheless, I learned a great deal, both scientific and personal, about him from the biography. For those interested in learning of Wright's life and contributions, this is the place to start. The reprint collection contains several papers that are now very hard to obtain. Finally, Wright summarized his own scientific work as well as that of many others in the four-volume set (Wright 1968, 1969, 1977, 1978).

[2] For a brief obituary see my article, 'Sewall Wright (1889—1988)', *Genetics* **119** (1988), 1—4.

[3] Wright 1988.

[4] Two people who studied with both Wright and Muller reached opposite conclusions. James D. Watson called Muller the greatest geneticist of his generation; Carlos Offerman, in contrast, once told me that he placed Wright higher. Offerman was greatly impressed by the power of path analysis, which he said gave Wright the edge.

[5] These contributions are briefly summarized, along with a personalized, anecdotal description of Wright, in my article 'Sewall Wright, the Scientist and the Man', *Perspect. Biol. Med.* **25** (1982), 279—294 (reprinted in *Genetic Perspectives in Biology and Medicine*, ed. Edward D. Garber [Chicago: University of Chicago Press, 1985], pp. 121—136). See Provine, *Sewall Wright*, for a much fuller account.

[6] Wright's lectures and seminar talks typically went well past the scheduled time. The last time he taught a formal course there were two students, Yuichiro Hiraizumi and the late Ove Frydenberg; both became distinguished geneticists. The class met at 11:00 and inevitably ran overtime through the noon hour, creating hunger pangs until the students suggested that they routinely continue the class in a nearby drugstore booth.

[7] I also benefited from a Wright review: he refereed the first manuscript that I ever sent to *Genetics* and showed how one of my approximate formulae could be made exact.

[8] On one occasion a mathematical paper by Max Delbrück was sent to an anonymous reviewer, Kimball Atwood, who did a very thorough and careful review. Delbrück later wrote asking the editors to thank Sewall Wright for such an excellent review.

[9] This collaboration is discussed by William Provine in *Dobzhansky's Genetics of Natural Populations*, ed. Richard C. Lewontin, John A. Moore, William B. Provine, and Bruce Wallace (New York: Columbia University Press, 1981). See also Provine, *Sewall Wright*.

[10] Provine, *Sewall Wright*.

[11] The University of Wisconsin had no facilities for rearing guinea pigs. I believe this was fortunate, for the guinea pig work was very time-consuming. The freed time gave Wright the chance to analyze his accumulated data and write his four books.

[12] Wright knew how to extract cube roots before he started school. On his first day in class, the astonished first-grade teacher had him demonstrate this skill for the eighth-grade class. According to Wright, this act brought him instant unpopularity among the other students, and he decided never again to volunteer anything in class.

[13] Curiously, Haldane also used a period of forced inactivity caused by a war injury to read the same book, P. G. Tait's *Elementary Treatise on Quaternions*. As far as I know, neither he nor Wright ever used the technique in his research. Like Haldane and Fisher, Wright was interested in astronomy. His illness didn't deter him from climbing to the roof of his railroad car to view Halley's comet.

[14] Wright 1917.

[15] Wright 1920a, 1926.

[16] Wright 1921a.

[17] Wright 1925. It has been remarked that this was the best thing to come from the Harding administration.

[18] Ching Chun Li, *An Introduction to Population Genetics* (Peking: National Peking University Press, 1948). A revised and expanded edition was published in 1955 by the University of Chicago Press.

[19] On the occasion of Wright's ninetieth birthday banquet, a Wisconsin sociologist said that Wright's contributions to the social sciences were such that he was prepared to offer him an assistant professorship. Wright's absence of horn-tooting is brought out by the fact that younger social scientists at Wisconsin more than once were surprised to learn that the inventor of path analysis was not only still alive, but at their own institution.

[20] Wright 1983.

[21] Wright 1922b.

[22] As evidence that inbreeding doesn't necessarily have harmful effects on vigor, longevity, and performance, I'll note that Wright's parents were cousins.

[23] Wright 1920b. The price of this 67-page bulletin was 15 cents.

[24] Wright 1922a.

[25] Jay L. Lush, *Animal Breeding Plans*, 3rd ed. (Ames: Iowa State College Press, 1945).

[26] Arthur B. Chapman, ed., *General and Quantitative Genetics* (Amsterdam: Elsevier Science Publishers, 1985); see esp. chap. 10.

[27] I have often wondered how the history of American genetics would have been changed had Wright and Little arrived at Harvard in reverse order. Little went on to develop inbred mouse strains and, through persistence, perspicacity, and personality, founded the Jackson Laboratory in Maine. This certainly would not have happened had Wright gone into mouse genetics; he was no promoter. On the other hand, Wright was a much deeper thinker than Little, and undoubtedly the physiological genetics of the mouse would have developed more rigorously than it did.

[28] Wright 1917—18.

[29] Wright 1918.

[30] H. J. Muller, 'Further Studies on the Nature and Causes of Gene Mutations', *Proc. VI Int. Congr. Genet., 1* (1932), 213—255.

[31] Wright 1941.

[32] J. B. S. Haldane, *The Causes of Evolution* (New York: Harper Brothers, 1932; reprinted Ithaca: Cornell University Press, 1966).

[33] J. B. S. Haldane, 'The Effect of Variation on Fitness', *Amer. Nat.* 71 (1937), 337—341; *idem*, 'The Cost of Natural Selection', *J. Genet.* 55 (1957), 511—524.

[34] R. A. Fisher, *The Genetical Theory of Natural Selection* (Oxford: Clarendon, 1930; revised ed. New York: Dover, 1958).

[35] R. A. Fisher, 'The Correlation between Relatives on the Supposition of Mendelian Inheritance', *Trans. Roy. Soc. Edinburgh* **52** (1918), 399—433. Just as Wright had trouble getting his corn-hog correlation paper published, Fisher had trouble with this one. It was rejected around 1916 by the Royal Society of London, and it is said that, in their lack of enthusiasm for this paper, the two reviewers, R. C. Punnett and Karl Pearson, agreed for the first and only time. The paper was later published in Edinburgh, but only after Fisher's friend and benefactor, Major Leonard Darwin, provided a subsidy. For a history of the bitter controversy between Mendelists and biometricians, in which Punnett and Pearson played prominent roles, see W. J. Provine, *Origins of Theoretical Population Genetics* (Chicago: University of Chicago Press, 1971).

[36] Motoo Kimura, 'On the Change of Population Fitness by Natural Selection', *Heredity* **12** (1958), 145—167. Fisher never regarded this paper as going beyond his own work, and as a condition for its publication — he was then editor of *Heredity* — he required a footnote to this effect.

[37] Wright 1951.

[38] Wright 1945. I find it interesting that Wright's later papers, although the equations become more general and comprehensive, are easier to understand than the earlier ones. I think it was Poincaré who said that the general formulation is often simpler than special cases.

[39] Wright 1942, 1949.

[40] T. Nagylaki, 'Gustave Malécot and the Transition from Classical to Modern Population Genetics', *Genetics* **122** (1989), 253—268.

[41] Wright would probably have been empathetic to a statement by Einstein, who is said to have remarked that since the mathematicians had invaded the field he could no longer understand relativity.

[42] See Lewontin *et al.*, *Dobzhansky's Genetics*; Wright's papers with Dobzhansky are reprinted here, along with a discussion of their place in the history of population genetics.

[43] Wright 1968, 1969, 1977, 1978.

[44] I find it ironic that some of Wright's most enthusiastic disciples, Th. Dobzhansky and Julian Huxley being the two most prominent, had very little understanding of his work. Their talents were biological and linguistic, not mathematical.

[45] Wright 1978.

[46] Wright 1931. He also published a less technical account in 1932. See his 1988 paper for a restatement and defense of this presentation.

[47] Wright 1982. Provine in *Sewall Wright* has called attention to changes over the years in Wright's statements about random gene frequency drift and selection. To me, however, a more striking feature is the consistency and single-minded determination with which he upheld this view; it has remained essentially the same since its inception.

[48] Provine has been critical of Wright's adaptive surface diagrams, and indeed Wright was not consistent in labeling the abscissas: sometimes they were genotypes, sometimes genotype frequencies, and sometimes gene frequencies. Only the last gives a continuous fitness surface in a finite population, and this is what I have used. The mean fitness of a

population is not a function of the gene frequencies alone, but with constant genotypic fitnesses, random mating, and unlinked or loosely linked loci this is true to a satisfactory approximation. Wright always emphasized that his diagrams were metaphorical and not intended to have a precise meaning.

[49] Wright 1935.

[50] Wright refers to the third phase as group selection, but he makes clear that he is thinking mainly of emigration from the favored group. For example, he says: "By expansion of numbers and excess migration such races tend to bring the species as a whole under control of this peak. Intergroup selection of·this sort [my italics], with respect to racial differentiation that has jointly adaptive and non-adaptive aspects, seems to provide the most effective mechanism for testing many alleles at each locus and many combinations of these and is thus the most effective mechanism for a continuing evolutionary process" (Wright 1942: 244). This is not group selection as many evolutionists and animal behaviorists use the word, and this has been a source of confusion.

[51] J. B. S. Haldane, 'Natural Selection', in *Darwin's Biological Work: Some Aspects Reconsidered*, ed. P., R. Bell (New York: Wiley, 1959), pp. 140—141.

[52] J. F. Crow, W. R. Engels, and C. Denniston, 'Phase Three of Wright's Shifting Balance Theory', *Evolution* 44 (1990), 233—247.

[53] This is based on several personal conversations with Fisher, who often visited the University of Wisconsin in the latter years of his life. He did not visit Wright, however. For a discussion of the unpleasant personal relationship between these two men, see Provine, *Sewall Wright*; part of the Fisher viewpoint on adaptive surfaces is given in a letter from Fisher to Wright, quoted on p. 274.

[54] Stephen Jay Gould, 'The Return of Hopeful Monsters', *Nat. Hist.* 86 (June—July 1977), 22—30; S. J. Gould and Niles Eldredge, 'Punctuated Equilibria: The Tempo and Mode of Evolution Reconsidered', *Paleobiology* 3 (1977), 115—151.

[55] Motoo Kimura, *The Neutral Theory of Molecular Evolution* (Cambridge: Cambridge University Press, 1983).

[56] N. H. Barton and S. Rouhani, 'The Frequency of Shifts between Alternative Equilibria', *J. Theoret. Biol.* 125 (1987), 397—418; the bibliography includes other relevant work.

[57] Wright 1953. I attended this lecture and recall the surprise and bewilderment it elicited among my co-attendees.

[58] Wright 1953: 6. Subsequent quotations are from Wright 1964. See also Wright 1975.

[59] Marvin Minsky, *The Society of Mind* (New York: Simon and Schuster, 1985), p. 322.

[60] Wright 1964.

[61] H. Kacser and J. A. Burns, 'The Molecular Basis of Dominance', *Genetics* 97 (1981), 639—666.

[62] See Nagylaki, 'Gustave Malécot' (above, n. 40).

WRIGHT LITERATURE CITED

This list includes only those Wright articles referred to in the text and footnotes. The first reference includes a complete list of his publications.

Collected Reprints

1986. *Evolution, Selected Papers*. Edited with introductory materials by William B. Provine. Chicago: University of Chicago Press.

Books

1968. *Evolution and the Genetics of Populations. I. Genetic and Biometric Foundations.* Chicago: University of Chicago Press.

1969. *Evolution and the Genetics of Populations. II. The Theory of Gene Frequencies.* Chicago: University of Chicago Press.

1977. *Evolution and the Genetics of Populations. III. Experimental Results and Evolutionary Deductions.* Chicago: University of Chicago Press.

1978. *Evolution and the Genetics of Populations. IV. Variability Within and Among Natural Populations.* Chicago: University of Chicago Press.

Articles

Papers included in *Evolution, Selected Papers* are marked by an asterisk.

1917. 'The Average Correlation within Subgroups of a Population', *J. Wash. Acad. Sci.* **7**: 532—535.

1917—18. 'Color Inheritance in Mammals', *J. Hered.* **8**: 224—235, 373—378, 426—430, 473—475, 476—480, 521—527, 561—564; **9**: 33—38, 89—90, 139—144, 227—240.

1918. 'On the Nature of Size Factors', *Genetics* **3**: 367—374.

1920a. 'The Relative Importance of Heredity and Environment in Determining the Piebald Pattern of Guinea Pigs', *Proc. Nat. Acad. Sci.* **6**: 320—332.

1920b. 'Principles of Livestock Breeding', Bull. no. 905, U.S. Dept. Agric.

1921a. 'Correlation and Causation', *J. Agric. Res.* **20**: 557—585.

1921b. 'Systems of Mating', *Genetics* **6**: 111—123, 124—143, 144—161, 162—166, 168—178.

1922a. 'Coefficients of Inbreeding and Relationship', *Amer. Nat.* **56**: 330—338.

1922b. 'The Effects of Inbreeding and Crossbreeding on Guinea Pigs', Bull. no. 1090, U.S. Dept. Agric.

1925. 'Corn and Hog Correlations', Bull. no. 1300, U.S. Dept. Agric.

1926. 'A Frequency Cruve Adapted to Variation in Percentage Occurrence', *J. Amer. Statist. Assoc.* **21**: 161—178.

1931*. 'Evolution in Mendelian Populations', *Genetics* **16**: 97—159.

1932*. 'The Roles of Mutation, Inbreeding, Crossbreeding and Selection in Evolution', *Proc. VI Int. Congr. Genet.* **1**: 356—366.

1935*. 'The Analysis of Variance and the Correlations between Relatives with Respect to Deviations from an Optimum', *J. Genet.* **30**: 243—256.

1941. 'The Physiology of the Gene', *Physiol. Rev.* **21**: 487—527.

1942*. 'Statistical Genetics and Evolution', *Bull. Amer. Math. Soc.* **48**: 223—246.

1945*. 'The Differential Equation of the Distribution of Gene Frequencies', *Proc. Nat. Acad. Sci.* **31**: 383—389.

1949. 'Adaptation and Selection', in *Genetics, Paleontology and Evolution*, ed. Glenn L. Jepson, Ernst Mayr, and George Gaylord Simpson, pp. 365—389. Princeton: Princeton University Press, 1949.

1951*. 'The Genetical Structure of Populations', *Ann. Eugen.* **15**: 323—354.

1953. 'Gene and Organism', *Amer. Nat.* **87**: 5—18.

1964. 'Biology and the Philosophy of Science', *Monist* **48**: 265—290.

1975. 'Panpsychism and Science', in *Mind in Nature*, ed. David R. Griffen and John E. Cobb, pp. 79—88. Washington: University Press of America.

1978*. 'The Relation of Livestock Breeding to Theories of Evolution', *J. Anim. Sci.* **46**: 1192—1200.

1982*. 'Character Change, Speciation, and the Higher Taxa', *Evolution* **36**: 427—443.

1983. 'On "Path Analysis in Genetic Epidemiology: A Critique," ' *Amer. J. Human Genet.* **35**: 757—768.

1988. 'Surfaces of Selective Value', *Amer. Nat.* **131**: 115—123.

WILLIAM B. PROVINE

THE R. A. FISHER—SEWALL
WRIGHT CONTROVERSY*

INTRODUCTION

The intense controversy between R. A. Fisher and Sewall Wright, which lasted from 1929 until Fisher's death in 1962, was both highly visible and very influential in modern evolutionary biology. The controversy between Fisher and Wright has had a more fundamental and lasting impact upon evolutionary biology than any other controversy or rivalry in this century. Indeed, the controversy did not end with Fisher's death. Since 1962, Wright has produced a steady stream of papers and his four volume *Evolution and the Genetics of Populations* (1968, 1969, 1977, 1978a) in which he constantly contrasts his interpretations with those of Fisher. Many others have joined the battle on both sides.

The aim of this paper is to elucidate the crucial issues in the Fisher—Wright controversy and to sort them out from the huge number of anecdotal stories about the controversy. I also will show how the controversy exerted such a major influence upon evolutionary theory and field research. I have drawn freely from my forthcoming biography of Sewall Wright (Provine 1986), to which the reader is referred for greater detail. The complete correspondence between Fisher and Wright is reproduced in the biography; an excellent selection of the correspondence with critical commentary may be found in Bennett (1983). All of Wright's correspondence will soon be available for scholarly investigation at the Library of the American Philosophical Society in Philadelphia, as Fisher's is already available at the University of Adelaide.

FIRST ACQUAINTANCE

Fisher and Wright first met in the summer of 1924. Fisher had read and been impressed by Wright's series of papers, 'Systems of Mating' (1921a—e) in which Wright had applied his method of path coefficients to the problem of inbreeding and its implications for breeding theory and evolution (the method of path coefficients is a technique for

201

Sahotra Sarkar (ed.), The Founders of Evolutionary Genetics, 201—229.
© 1992 *Kluwer Academic Publishers. Printed in the Netherlands.*

quantifying a given path of causation, from cause to effect, in a complex causal scheme; it is particularly effective when applied to definite causal lines and linear relations as in Mendelian inheritance). Fisher attended the International Mathematical Congress in Toronto in the summer of 1924, after which he travelled to Washington DC where Wright was working at the Animal Husbandry Division of the United States Department of Agriculture. Wright was aware of Fisher's 1918 paper, 'The Correlation between Relatives on the Supposition of Mendelian Inheritance', and his reputation as a bio-statistician. At the time Fisher was working at the Rothamsted Experimental Station as a statistician for plant breeders and Wright was serving in a similar role for animal breeders at the USDA. So they had a great deal to share and were eager to meet each other.

Wright and Fisher had a long conversation about animal and plant breeding, statistical techniques, and the quantitative consequences of Mendelian heredity. Wright gave Fisher copies of several of his papers on path coefficients and animal breeding, and in return Fisher promised to send Wright a copy of his 1922 paper, 'On the Dominance Ratio', which Wright had never seen. The paper arrived on schedule and Wright found it very stimulating. This initial interchange between Wright and Fisher started a chain of events that led from occasional exchange of Christmas cards to, beginning in 1929, serious correspondence about quantitative evolutionary theory.

FISHER, WRIGHT, AND MATHEMATICAL POPULATION GENETICS

Contrary to the opinion of many evolutionists, the Fisher—Wright controversy did not stem in any substantive way from disagreements in mathematical population genetics. On all occasions when Fisher and Wright appeared to disagree on quantitative questions, they were able to settle the differences and reach near total agreement.

Fisher and Wright used different quantitative approaches in their mathematical models of evolutionary change. Fisher favoured the differential and integral calculus, using continuous functions to build his models. Wright, however, used his method of path coefficients, which was particularly well adapted to the calculation of the effects of inbreeding in a finite population. Moreover, the quantitative approach of each was well suited to the theory of evolution each advocated. Thus,

Fisher's approach was perfect for developing his fundamental theorem of natural selection and Wright's approach for his shifting balance theory of evolution in nature, in which inbreeding played a major role in the production of heritable variation between populations.

There can be no doubt that the same data sets can be interpreted very differently by using different quantitative approaches. The intense arguments between statisticians and biometricians over the best quantitative techniques to use in particular circumstances illustrates this point. Yet in the case of controversies between Fisher and Wright, the differences in their quantitative methods do not appear to explain their fundamental differences on questions concerning the evolutionary process. Their different quantitative methods led to almost exactly the same numerical results regarding the quantitative effects of inbreeding, mutation, migration, selection, and other variables, upon the statistical distribution of genes in a population.

DIFFERENCES OF THE EARLY YEARS

Fisher was the first to have the idea of modelling the evolutionary process upon the changes in a statistical distribution of genes in a population. This was his great invention in the important mistitled 1922 paper, 'On the Dominance Ratio' (Fisher 1922). In his now famous 1918 paper, Fisher had examined the statistical consequences of such variables as genic interaction (epistasis), dominance, assortative mating, multiple alleles, and linkage upon the correlations between relatives. In the 1922 paper he extended the analysis to examine the effects of these and other variables upon the statistical distribution of genes in the population. Fisher was proposing that the changes in this distribution constituted the evolutionary history of the population.

The fundamental thoughts underlying many of the later quantitative works of Fisher and Wright on evolution were contained in this seminal 1922 paper. Beginning with the Hardy—Weinberg equilibrium, the distribution of genes expected under ideal assumptions, Fisher analysed the influence of selection, dominance, heterozygote advantage, mutation rate, random extinction of genes, and assortative mating upon the distribution of genes in the population. He showed, for example, that at a single locus with two alleles, selection favouring one homozygote would lead to the elimination of the other allele. If, on the other hand, selection favoured the heterozygote, then the result was a stable equilibrium of the distribution of the alleles in the population. He also demonstrated that the survival at low frequency in a population of a rare mutation depended more upon chance than selection. Thus, large populations, in which a mutation would occur more often in absolute numbers, would be expected to maintain a higher genetic variability than small populations. This was a very exciting paper for those few who could understand what Fisher had done.

Wright did not have an extensive background in mathematics, not even in differential and integral calculus. He could not follow all of Fisher's differential equations and derivations in the 1922 paper, but he certainly did understand Fisher's attempt to incorporate the effects of crucial variables into one statistical distribution of genes in a population. Indeed, Wright was so impressed that within a year he was hard at work developing his own version of the statistical distribution of genes, using his method of path coefficients rather than the differential equations used by Fisher.

As Wright developed his model, he discovered that it differed in some significant and puzzling ways from that of Fisher. When Wright used his method of path coefficients to calculate the rate of decrease of heterozygosis in a population under no selection or mutation he obtained the figure $1/2N$, where N was the effective population size. Using differential equations, Fisher had arrived at the corresponding figure for the rate of decay of allele frequency as $1/4N$ (Fisher 1922, p. 330). Also, the factors in Wright's distribution for selection and dominance did not agree with Fisher's model, although Wright was not confident in these cases that his own method led to reliable results. Wright wrote up his paper on evolution and had it typed in the late fall of 1925, just as he was preparing to leave the USDA for a professorship at the University of Chicago. Despite his strong desire to submit the paper for publication, Wright held back because of the discrepancies with Fisher's model and his hope to push beyond it in some significant ways.

For the first 2 years at the University of Chicago, Wright was inundated with teaching duties and had little time to sort out the differences between his model and Fisher's. When he did attempt this on a few occasions, he could not work out the reasons for the differences in the models. Then, in 1928, Fisher published two papers on the evolution of dominance (Fisher 1928a,b). For reasons to be discussed later in this paper, Wright disagreed strongly with Fisher's theory of the evolution of dominance and wrote a rebuttal (Wright 1929a). Fisher thought that Wright disagreed with his calculations of the effects of small selection rates on gene frequencies over long periods and wrote a letter to Wright that said in part, "What I mainly want to know . . . is whether you agree with me that a very slight selective effect acting for a correspondingly long time will be equivalent to a much greater effect acting for a proportionately shorter time" (Fisher to Wright 6 June 1929). Wright replied that he was not challenging Fisher's mathematical calculation of the effects of selection in changing gene frequencies, but instead his conception of the evolutionary process (Wright to Fisher, 28 June 1929). Thus, in his published rebuttal to Wright's 1929 paper cited above, Fisher began by stating that Wright's "primary formulas differ in no essential respects from my own and that the selective intensity which inclines Professor Wright to reject the theory is in fact the same that originally led me to adopt it" (Fisher 1929, p. 553). Indeed, on the one quantitative issue where they did disagree (the value of the selection pressure upon the heterozygote), Fisher was able to show that Wright's calculations were incorrect, but when corrected were in full agreement with his own calculations. In other words, they had no disagreement whatsoever on the quantitative models themselves with regard to the evolution of dominance.

By August 1929, Wright had rewritten his paper on evolution and he sent a copy to Fisher with the explicit hope that Fisher could discover the reason for the discrepancy in their figures for the rate of decay of heterozygosis in finite populations ($1/4N$ for Fisher, $1/2N$ for Wright). Fisher found the mistake in his earlier derivation, thanking Wright profusely in a letter for pointing out the discrepancy and adding that "with this correction I find myself in entire agreement with your value $2N$ for the time of relaxation and with your corrected distribution for factors in the absence of selection" (Fisher to Wright, 15 October 1929). Fisher entered these changes in proof of his forthcoming book *The Genetical Theory of Natural Selection* (Fisher 1930).

However, their terms for selection in the stochastic distribution were still not identical. Wright noticed this particularly when Fisher sent him a copy of the book. Wright wrote to Fisher asking for a more detailed derivation of his selection term, a request to which Fisher responded immediately. With Fisher's derivation in hand, Wright was able to rederive his own selection term so that it gave results identical with those of Fisher. Wright entered this new derivation in both the proofs of his review of Fisher's book for the *Journal of Heredity* (Wright 1930) and in the proofs of his paper on evolution then in press at *Genetics* (Wright 1931). When Wright wrote to Fisher to tell him of these developments, Fisher responded, "I am glad to hear that the little discrepancies are clearing themselves up" (Fisher to Wright, 25 October 1930). In his review of the *Genetical Theory of Natural Selection*, after an account of his correspondence with Fisher, Wright stated clearly that "our mathematical results on the distribution of gene frequencies are now in complete agreement as far as comparable, although based on very different methods of attack" (Wright 1930, p. 352).

Only once after this full agreement of quantitative results in the early 1930s did Fisher and Wright appear to reach different conclusions as a consequence of different quantitative approaches. This concerned their disagreement about evolution in the rare species, *Oenothera organensis*, found only in the Organ Mountains of southern New Mexico [for an account of this controversy, see Provine (1986, Chapter 13)]. In response to Wright's analysis of the data (Wright 1939), Fisher responded in the second edition of *Genetical Theory of Natural Selection* (1958) with his own analysis using differential equations. He concluded that his quantitative results were 'very different' from those of Wright in 1939, and he attributed the differences to Wright's 'failure to develop any explicit formulae' and his reliance upon 'extensive numerical calculations based upon trial values of numerous constants he introduces' (Fisher 1958, p. 109). Wright answered with a careful comparison of the practical consequences of his and Fisher's models, concluding convincingly that they were virtually identical.

Thus, the differences in quantitative approaches of Fisher and Wright did not entail significant differences in their evolutionary theories. Nor was the tension between their quantitative approaches a creative force in evolutionary biology. Most modern evolutionists use whatever quantitative approach best suits the needs of the project at hand, and switch back and forth freely between Fisherian and Wrightian models. Frequently, these switches may be found in a single paper or chapter. The really important and influential tensions between Fisher and Wright had almost nothing to do with differences in the mathematical methods each employed.

PERSONAL DIFFERENCES

Anywhere theoretical population genetics is discussed stories about the personality differences between Fisher and Wright abound. The rather few occasions when they were thrown together after the early 1930s, for example at Iowa State University, North Carolina State University, and the University of Cambridge, all generated gossip about how poorly they communicated in person. All correspondence between them, except for the unimpeded exchange of published papers, ceased in 1932. There were differences between them in both temperament and personality. In 1964, J. B. S. Haldane described these differences by saying that Fisher, when alive, "preferred attack to defense. Wright is one of the gentlest men I have ever met and if he defends himself, will not counter-attack" (Haldane 1964, p. 344). I think that Haldane's assessment, although reflecting widely held opinions of evolutionists about Fisher and Wright, was misleading at best. Fisher probably did prefer attack to defence, but he also could have a strong defensive reaction, particularly when he thought the criticism or attack upon him was wrong or unwarranted [see the biography of Fisher by his daughter. Joan Fisher Box (1978)]. Haldane's characterization of Wright was just plain wrong. Wright was very shy, but was not gentle with his critics. How Haldane could have believed that Wright would not counter-attack is a mystery to me. He was actually the inveterate master of the complete counter-attack, frequently repeated many times over. Wright has always fully answered his critics.

A conflict was bound to result from these differences in personality. Wright continued criticizing Fisher, particularly his theory of the evolution of dominance, well after Fisher thought he had rebutted Wright's argument and after he had suggested to Wright in a letter that they cease to publish on their differences on the evolution of dominance. In 1933, however, Wright felt that he had never had the opportunity to really express in print his whole theory of the evolution of dominance. Thus, when Wright's paper on the evolution of dominance appeared (1934a), one section of which repeated his critique of Fisher's theory, Fisher wrote a very angry rebuttal (1934), which was followed by a strongly worded reaction from Wright (1934b). They were never on friendly terms again.

Although these personal differences were real and were certainly related to the tone of their disagreements over the years, and added a certain theatrical air to the conflict, I do not think that these differences go far enough in explaining the really important aspects of their conflict. Important scientists often have conflicts of personality, but such conflicts do not frequently have an important effect upon the advancement of the field. Even if Fisher and Wright had been the best of personal friends and had disagreed amicably, as many friends do, their disagreements on evolutionary theory would have been substantive and influential.

EVOLUTIONARY THEORY

Fisher's Theory

Qualitative evolutionary theory was the subject on which Fisher and Wright disagreed so strongly, and to such great effect. From his early student days at Cambridge University, Fisher was a devoted and self-conscious neo-Darwinian. By this, I mean that his general view of the evolutionary process fit closely with that of E. B. Poulton, E. Ray Lankester, Raphael Meldola, Karl Jordan, Leonard Darwin, and others who believed that natural selection was at all levels by far the most pervasive and important mechanism of evolution in nature. Under this view, every character of an organism had been shaped by the action of natural selection, whether or not biologists of the day had been able to spot the utility of the character. Combined with Mendelian heredity, understanding of the origin of variation, and recombination, Darwin's theory of natural selection was to Fisher's mind the nearly complete mechanism of evolution in nature.

Fisher was also greatly interested in mathematics, astronomy, and physics, and he found especially attractive the deep explanatory power of simple quantitative laws such as the inverse square law of gravitational attraction, Boyle's gas laws, and the second law of thermodynamics. He wanted to find the correspondingly simple quantitative law that would allow evolutionary phenomena to fall in place. His first attempt in this direction was in his 1922 paper, 'On the Dominance Ratio', in which he argued that an equation representing the stochastic distribution of Mendelian determinants in a population over time was the key to an accurate and quantitative understanding of evolution in that population.

Among many simplifying assumptions in the 1922 paper, Fisher assumed his population was extremely large and consequently had high storage of genetic variability. In such a population, he was able to demonstrate that his stochastic distribution led to the certain conclusion that of all the variables (selection, dominance, mutation rate, random extinction of genes, assortative mating, and epistasis), natural selection acting upon single genes was the supreme determinant of the evolutionary process. A mutation rate far higher than any observed in nature could be balanced by a tiny selection rate against it. Epistasis and random extinction were negligible. Fisher likened the stochastic distribution of genes, dominated by natural selection, to the general laws of the behaviour of gases.

The investigation of natural selection may be compared to the analytic treatment of the Theory of Gases, in which it is possible to make the most varied assumptions as to the accidental circumstances, and even the essential nature of the individual molecules, and yet to develop the general laws as to the behaviour of gases, leaving but a few fundamental constants to be determined by experiment (1922, pp. 321−2).

However, the whole stochastic distribution was a rather messy equation, especially as more and more variables were incorporated into it. Thus, by 1930 Fisher had reached the view that the law was the real counterpart of the great laws of the physical sciences was his "fundamental theorem of natural selection", which stated in words said, "The rate of increase in fitness of any organism at any time is equal to its genetic variance in fitness at that time" (Fisher 1930, p. 35). After deriving and explaining the theorem, Fisher wrote:

Professor Eddington has recently remarked that 'The law that entropy always increases − the second law of thermodynamics − holds, I think, the supreme position among the laws of nature'. It is not a little instructive that so similar a law should hold the supreme position among the biological sciences (Fisher 1930, pp. 36−7).

A fundamental aspect of Fisher's approach to biology was his expectation that the 'laws' of biology would be similar to the laws of physics.

Wright's Theory

Wright's theory of evolution in nature was far more strongly modelled upon biology then was Fisher's. Wright built directly upon his extensive knowledge of experimental laboratory genetics, and the experience of animal and plant breeders. Wright also drew upon and was influenced by a very different natural history tradition than the neo-Darwinian one with which Fisher associated himself.

Four major research projects were influential in shaping Wright's theory of evolution in nature: (1) William Castle's selection experiment upon hooded rats; (2) Wright's thesis research on the interaction effects in guinea-pigs of the Mendelian factors determining colour and hair direction; (3) Wright's work on inbreeding, outbreeding, and selection in guinea-pigs; and (4) the analysis of the transformation of the Shorthorn breed of cattle during its foundation period (Wright 1978b).

From Castle's selection experiment upon hooded rats Wright learned two very important, but quite different points: that direct mass selection (selection in a random breeding population) was a powerful means of genetically changing the expression of a character, and that mass selection also had built-in limitations. The limitations had long been known to professional animal and plant breeders. Severe mass selection could indeed rapidly change the expression of a character such as milk production, but at the extreme and often unacceptable cost of deleterious side effects, distressingly expressed to breeders as loss of fecundity. In the breeding of large animals such as cattle or horses, mass selection was a slow and frustrating process, especially when the characters being selected had a low heritability (as was frequently the case). Fisher was also much impressed by Castle's selection experiments with hooded rats and cited this work frequently, especially in connection with his theory of the evolution of dominance. But Fisher emphasized only the positive effects of mass selection as had Darwin, a rather curious view considering that Darwin was highly knowledgeable of both animal and plant breeding, and Fisher was working at the Rothamsted Station at the time he published his theory of the evolution of dominance. If anything, Wright was more impressed by the limitations of mass selection.

Wright's thesis research upon the interaction effects of colour factors in guinea-pigs demonstrated that organisms were built up of complex interaction systems of genes rather than being, as Wright frequently emphasized, a mere mosaic of unit characters each determined by a single gene. From this fact of interaction, Wright deduced that to the breeder or under natural selection the selective process would be most effective if it operated upon interactive systems of genes rather than upon single genes. In a large random breeding population, the possible distinctive interaction systems of genes were rarely phenotypically expressed and thus not exposed to selection. As Fisher had shown clearly, in a large random breeding population selection was effectively limited to mass selection of single genes. Wright believed that both breeders and natural selection had more effective mechanisms than mere mass selection.

A clue to this more effective mechanism came from Wright's work on highly inbred strains of guinea-pigs at USDA. Because of the random fixation of genes caused by the many generations of inbreeding (brother-sister mating), each strain or family became fixed with a highly distinctive (almost) homozygous genetic complement. The breeding process had revealed the interaction systems so well hidden in the random breeding control population making them available for the selection process. Of course, breeders would have to use intermediate rather than severe inbreeding to avoid the general decline in vigour and fecundity that usually followed intense inbreeding.

From his analysis of the origin of the Shorthorn breed of cattle, Wright discovered that breed had indeed undergone a time of intense inbreeding during its foundation period. Selection was applied to the variability revealed by the inbreeding. Diffusion from a very few herds to many others was accomplished by use of only a relatively few closely related sires, thus making over the entire breed. Wright thought that mass selection had played a relatively minor role.

Reasoning from his theory of optimal animal breeding to his theory of evolution in nature (as had Charles Darwin), Wright deduced that for natural selection to be a really efficient process, populations in nature must be subdivided into partially isolated subgroups small enough to cause the random drifting of genes and consequent manifestation of many different interaction systems. Optimally, the subgroups were large enough to prevent direct fixation by random draft, because this would lead to degeneration and extinction. Mass selection within subgroups followed by selective diffusion from particularly successful subgroups were the steps required for the transformation of the whole species.

Importance of Traditions in Natural History and Taxonomy

The other factor that must be kept in mind for understanding Wright's early expressions of his theory of evolution in nature is the natural history and taxonomic tradition that he followed, in contrast to the neo-Darwinian, wholly adaptationist tradition that Fisher followed. Right in Darwin's own work, there was a basic tension between adaptive and non-adaptive mechanisms of the evolutionary process (for explanation and documentation see Provine 1985a). Both traditions were very active and generating controversy in the last two decades of the nineteenth century and first two decades of the twentieth century. By the time Wright was formulating his theory of evolution in nature, there was a very strong tradition in both taxonomy and natural history that challenged the neo-Darwinian view. Represented by such figures as Moritz Wagner, John T. Gulick, David Starr Jordan, G. J. Romanes, Vernon L. Kellogg, H. E. Crampton, F. B. Sumner, A. C. Kinsey, G. C. Robson, and O. W. Richards, followers of this tradition argued that many, if not most, of the differences observed between closely-rated species were of no adaptive value whatsoever and such differences must have arisen from some other mechanism than natural selection. Geographical isolation was always raised as a contributing factor to this non-adaptive differentiation.

Wright clearly believed that his theory of population subdivision could explain both the adaptive and the non-adaptive differences, depending upon the degree of subdivision and consequent effect of random genetic drift, whereas the neo-Darwinian view could explain non-adaptive differences only by the rather weak explanation that such characters were linked with others of such high adaptive value that their combination was adaptive (the theory of correlation).

Population structure and traditions in taxonomy and natural history are therefore the keys to understanding the basic differences between the early evolutionary views of Fisher and Wright. Fisher's view, based upon very little evidence, was that evolution proceeded by the mass selection of single genes in very large random breeding populations. This view of the evolutionary process was impossible if natural populations were subdivided the way that Wright thought. Wright's belief, also based upon very little evidence from natural populations, was that the assumption of large random breeding population was unwarranted and that appropriately subdivided populations could lead to a better understanding of both artificial and natural selection. But Wright's view of the evolutionary process was impossible if Fisher was correct in his assumption about natural populations being large and random breeding. Although Wright could have no quarrel with Fisher's fundamental theorem of natural selection given Fisher's assumptions, he did think that the theorem, to be accurate, should be stated as follows: "The rate of increase in fitness of any population at any time is equal to its genetic variance in fitness at that time, *except as affected by mutation, migra-*

tion, change of environment, and effects of random sampling" (Wright to Fisher, 3 February 1931). The tension between these views was deep and inevitable, and moreover could not be settled by any amount of theorizing, no matter how quantitative. Only careful study of natural populations could resolve the tension.

THE CONTROVERSIES

In the space available here I will be able to examine briefly only three of the many controversies in which Fisher and Wright engaged, but these should give sufficient insight to see their significance. These three controversies were over: (1) the evolution of dominance, (2) the general mechanism of evolution in nature, and (3) evolution in the moth *Panaxia dominula*.

The Evolution of Dominance

Fisher applied his basic evolutionary theory first to the evolution of mimicry (Fisher 1927) and then to the evolution of dominance (Fisher 1928a, b). His thesis was that these apparent cases of evolution by discontinuous leap were in fact the result of deterministic small selection pressures acting over long periods of time upon very small heritable modifiers always available in large random breeding populations. Noting that most observed mutations in *Drosophila melanogaster* were wholly recessive, Fisher offered the following explanation. Most mutations, he argued, were recurrent, deleterious, and occurred at a finite rate. The mutations that geneticists had observed in the laboratory in *Drosophila* must also have occurred in nature. Since natural populations were large and random breeding, a deleterious mutant allele was likely to become fixed by reaching a state of equilibrium between adverse selection and recurrent mutation. Natural selection would tend to make the heterozygote and mutant homozygote phenotypically resemble the homozygous wild type. Castle's experiment on hooded rats had shown how selection could accumulate modifiers of a mutant to change its appearance very significantly. Thus dominance was not an immutable property of the gene. Because heterozygotes were vastly more numerous in the population than the mutant homozygotes, natural selection would over time make the heterozygotes resemble the homozygous wild type, thus accomplishing dominance. A recessive allele was a potential wild type allele and the force required for the change was a minute selective advantage. Fisher acknowledged that the selective advantages causing the evolution of dominance were extremely small: "It may be calculated that with mutation rates of the order of one in a million, the corresponding selection in the state of nature, though extremely slow, cannot be safely neglected in the case of the heterozygotes" (Fisher 1928a, p. 126).

Wright objected strongly to Fisher's theory of the evolution of dominance on two grounds (Wright 1929a). First, he objected to Fisher's whole theory of the mechanism of evolution in nature. He thought Fisher's theory ignored factors that, in his opinion, would swamp the tiny selection rates operating over long periods of time that were hypothesized in Fisher's theory of the evolution of dominance. Primary among these factors were random genetic drift from sampling effects in relatively small populations and the selective pressures caused by interactive effects of the rest of the genome. Second, he had reason to believe from his research in physiological genetics that recessivity generally resulted from inactivation of the gene, thus reducing the product produced by the dominant allele. Interestingly, in his first published reply to Fisher's theory of the evolution of dominance, Wright did not even mention small effective population sizes or random genetic drift. That was because he had not yet published his long paper on evolution in which he treated, in some detail, the issue of population size and random drift, and he did not see a reasonable way to introduce this complex issue in a brief note. Thus, in the first published interchange between Wright and Fisher, it was impossible to detect that what was really at issue was two basically different theories of evolution in nature.

In his published reply to Wright's criticism of his theory, Fisher even used selectively neutral genes to buttress his mass selectionist viewpoint:

As to ratios having neutral stability, there is one reason for thinking that the factors suffering the feeblest selective action will at any one time be the most numerous. The fact of those powerfully selected is quickly settled; they do not long contribute to the variance. It is the idlers that make the crowd, and very slight attractions may determine their drift. On the whole, it seems that the most reasonable assumption which we can make, on an obscure subject, is that the effect is approximately equal to the cause (Fisher 1929, p. 556).

Wright was genuinely surprised to find Fisher using the existence of nearly neutral genes to support his intensely selectionist view of the evolutionary process (what a contrast with Ohta and Kimura!). This time Wright felt compelled to use his arguments of effective population size and random drift in his published answer to Fisher, but when he sent the draft of this answer to Fisher, he also sent the manuscript of his long paper on evolution. Thus, in addition to wanting to clear up the inconsistencies in their quantitative models, Wright wanted to provide Fisher with the background to one of his strongest objections to Fisher's theory of the evolution of dominance.

In his published statement, Wright said that he could not understand how Fisher could use the existence of almost neutral factors to argue for the all prevailing power of natural selection. Indeed, Wright asked, how small did the effective breeding size (N) of a population have to be before random drift would swamp the effects of selection rates of the size postulated by Fisher?

Unfortunately it is difficult to estimate N in animal and plant populations. In the calculations, it refers to a population breeding at random, a condition not realized in natural species as wholes. In most cases, random interbreeding is more or less restricted to small localities. These and other conditions such as violent seasonal oscillation in numbers may well reduce N to moderate size, which for the present purpose may be taken as anything less than a million. If mutation rate is of the order of one in a million per locus, an interbreeding group of less than a million can show little effect of selection of the type which Dr. Fisher postulates even though there be no more important selection process and time be unlimited (Wright 1929b, p. 560).

Wright clearly believed that natural species were not random breeding populations. Instead, he believed that random breeding was 'more or less restricted to small localities' over the entire range of the species. Thus, the effective breeding size of a whole species was vastly less than the number of breeding individuals in the species.

Fisher's reply to this argument in a letter to Wright is very instructive.

I am not sure that I agree with you as to the magnitude of the population number N. To reduce it to the number in a district requires that there shall be *no* diffusions even over the number of generations considered. For the relevant purpose I believe N must usually be the total population on the planet, enumerated at sexual maturity, and at the minimum of the annual or other periodic fluctuation. For birds, twice the number of nests would be good. I am glad, however, that you stress the importance of this number (Fisher to Wright, 13 August 1929).

By this time, both Fisher and Wright knew that the real issue between them was not merely a disagreement about the evolution of dominance, but a deep disagreement about the evolutionary process in general.

Their disagreement over the evolution of dominance immediately became a subject of great interest to evolutionists. J. B. S. Haldane, H. J. Muller, C. R. Plunkett, E. M. East, and others immediately joined the controversy and interest in it has never died away. Almost every decade the question of the evolution of dominance is 'definitively' settled by one or another evolutionist, only to emerge again as a difficult issue. A good review up to 1978 may be found in Wright's *Evolution and the Genetics of Populations*, Volume 3 (1977), but much has appeared on the evolution of dominance since then.

General Theory of Evolution in Nature

The controversy over the evolution of dominance obviously raised for Fisher and Wright awareness of the differences in their general theories of the evolutionary process. However, widespread appreciation of the tension between their evolutionary theories came only with the publication of their major works between 1930 and 1932. Fisher's *Genetical Theory of Natural Selection* (appeared May 1930) was the first major work to explore in sophisticated quantitative detail the synthesis of Mendelian genetics with evolutionary theory and is a landmark in the history of twentieth century evolutionary biology. Here Fisher developed in detail his view that evolution in nature occurred in large random breeding populations in which the overwhelming factor determining changes in gene frequencies was natural selection acting upon individual genes. Wright's contrasting view of the evolutionary process can best be seen in his review of Fisher's book (Wright 1930), his big paper 'Evolution in Mendelian Populations' (Wright 1931), and in his paper delivered at the Sixth International Congress of Genetics, 'The Roles of Mutation, Inbreeding, Crossbreeding, and Selection in Evolution' (Wright 1932).

In the review Wright presented the contrasting view that he found most appealing. Instead of the large panmictic population emphasized by Fisher, he argued that,

A much more favourable condition would be that of a large population, broken up into imperfectly isolated local strains. . . . The rate of evolutionary change depends primarily on the balance between the effective size of population in the local strain (N) and the amount of interchange of individuals with the species as a whole (*m*) and is therefore not limited by mutation rates. The consequence would seem to be a rapid differentiation of local strains, in itself non-adaptive, but permitting selective increase or decrease of the numbers in different strains and thus leading to relatively rapid adaptive advance of the species as a whole. Thus, I would hold that a condition of subdivision of the species is important in evolution not merely as an occasional precursor of fission, but also as an essential factor in its evolution as a single group (Wright 1930, pp. 354–5).

This was the first clear statement of what Wright later termed his 'three phase shifting balance' theory of evolution, involving large subdivided populations, random drift, intrademe, and interdeme selection. This was a clear alternative to Fisher's whole theory of evolution in nature.

The general impression of the differences between the theories of Wright and Fisher did not, however, reflect the full sophistication of either theory. I do not think that the impact of Fisher's theory of the inevitable deterioration of the environment, which caused a continual change in fitnesses and therefore in the parameters of his fundamental theorem, was well appreciated in the early years. Nor was Wright's shifting balance theory widely understood or appreciated. Instead, for understandable reasons, the conflict between their evolutionary theories became seen in the 1930s and, for the most part, later as one between Fisher's pan-selectionism and Wright's random genetic drift. Wright himself initially neglected to take into account the deterioration of the environment in evaluating Fisher's theory, although he did in the 1932 paper after Fisher complained in a letter (Fisher to Wright, 19 January 1931). For his part, Fisher never relinquished the view that Wright was advocating the importance of straight random genetic drift as an important mechanism of evolution, whereas Wright always argued that in his shifting balance theory random drift merely provided the variation upon which natural selection then acted to provided adaptive advance.

To be sure, there was much room for confusion about Wright's shifting balance theory of evolution. Since the late 1940s, Wright has consistently denied that he ever attributed any important role to random drift, except as a mechanism for generating variability upon which selection then acted. Thus, in 1967 Wright stated:

Many critics have seized on the concept of random drift that was proposed and have asserted that I have advocated this as a significant *alternative* to natural selection.

Actually, I have never attributed any evolutionary significance to random drift except as a trigger that may release selection toward a higher selective peak through accidental crossing of a threshold (Wright 1967, p. 254–5).

And more recently in 1982, Wright declared: "I emphasize here that while I have attributed great importance to random drift in small local populations as providing material for natural selection among interaction systems, I have never attributed importance to non-adaptive differentiation of species" (Wright 1982, p. 12). These statements must be compared with what Wright actually said in the years 1929–1932.

(1) The non-adaptive nature of the differences which usually seem to characterize local races, subspecies, and even species of the same genus indicates that this factor of isolation is in fact of first importance in the evolutionary origin of such groups, a point on which field naturalists (e.g., Wagner, Gulick, Jordan, Osborn, and Crampton) have long insisted (Wright 1929b, pp. 560–1).

(2) The actual differences among natural geographical races and subspecies are to a large extent of the non-adaptive sort expected from random drifting apart (Wright 1931, p. 127).

(3) Fisher's theory is one of complete and direct control by natural selection while I attribute greatest immediate importance to the effects of incomplete isolation (Wright 1931, p. 149 fn.).

(4) The direction of evolution of the species as a whole will be closely responsive to the prevailing conditions, orthogenetic as long as these are constant, but changing with sufficiently long continued environmental change (Wright 1931, p. 151).

(5) Adaptive orthogenetic advances for moderate periods of geologic time, a winding course in the long run, non-adaptive branching following isolation as the usual mode of origin of subspecies, species, perhaps even genera, adaptive branching giving rise occasionally to species which may originate new families, orders, etc., . . . are all in harmony with this interpretation (Wright 1931, p. 153).

(6) Under the shifting balance process complete isolation originates new species differing for the most part in non-adaptive respects, but is capable of initiating an adaptive radiation as well as of parallel orthogenetic lines, in accordance with the conditions (Wright 1931, p. 158).

(7) Complete isolation of a portion of a species should result relatively rapidly in specific differentiation, and one that is not necessarily adaptive. The effective intergroup competition leading to adaptive advance may be between species rather than races. Such isolation is doubtless usually geographic in character at the outset, but may be clinched by the development of hybrid sterility (Wright 1932, p. 363).

(8) That evolution involves non-adaptive differentiation to a large extent at the subspecies and even the species level is indicated by the kinds of differences by which such groups are actually distinguished by systematists. It is only at the subfamily and family levels that clearcut adaptive differences become the rule (Robson 1928; Jacot 1932). The principal evolutionary mechanism in the origin of species must then be an essentially non-adaptive one (Wright 1932, pp. 363–4).

(9) Subdivision into numerous local races whose differences are largely non-adaptive has been recorded in other organisms wherever a sufficiently detailed study has been made. [There follows citation of the work of Gulick, Crampton, David Starr Jordan, Ruthven, Kellogg, Osgood, Kinsey, Osborn, Rensch, Schmidt, David Thompson, and Sumner](Wright 1932, pp. 364—5).

Viewed all together at one time, these citations illuminate the question of why Wright's shifting balance theory was so misunderstood in the 1930s and later. The careful reader of Wright's papers in 1932 would almost certainly conclude that non-adaptive random drift following isolation was a primary mechanism in the origin of races, subspecies, species, and perhaps genera. One can easily understand why Fisher and other biologists understood Wright to be saying that random drift was an important mechanism alternative to selection in the origin of subspecies and species. Yet, at the same time, one can understand why Wright insists with reason that he has always argued that evolution in nature depends upon a balance of forces, of which random drift is only one. In the early 1930s, however, Wright correctly understood the taxonomists and naturalists to be telling him that most of the differences between closely related species were non-adaptive. Thus, he set the 'balance' in his shifting balance theory to give room for differentiation at the species level from random drift. Later, after the 1940s, when systematists led by Ernst Mayr and David Lack argued that non-adaptive differences between species were rare and adaptive ones the rule, Wright naturally saw his shifting balance theory as applicable.

Controversy Over Panaxia Dominula

Ever since the beginning of evolutionary biology, conspicuous polymorphisms in natural populations had presented evolutionists with a serious problem. How could a single primary mechanism of evolution, whether natural selection, inheritance of acquired characters of even an orthogenetic force lead to conspicuous dimorphisms within a single interbreeding population? All naturalists were familiar with at least some cases of such dimorphism. An obvious, but important reason why the debates over explaining the origin of conspicuous polymorphisms have been so persistent is that, until the rise of molecular biology, conspicuous polymorphisms were the most easily accessible (often it seemed, the only) measurable heritable characteristics. Thus, conspicuous polymorphisms have been constantly in the forefront of research on evolution in natural populations.

Darwin concluded in the Origin that natural selection could not be the explanation of polymorphic species (Darwin 1859). Indeed, when he defined the concept of natural selection in chapter IV of the Origin, Darwin specifically dissociated natural selection from polymorphism:

This preservation of favourable variations and the rejection of injurious variations, I call Natural Selection. Variations neither useful nor injurious would not be affected by natural selection, and would be left a fluctuating element, as perhaps we see in the species called polymorphic (Darwin 1859, p. 81).

Conspicuous polymorphisms were the focus of disagreement about adaptive versus non-adaptive evolution from Darwin's day until the 1940s, when the prevailing view of evolutionists became strongly adaptationist. But before that, the prevailing view was almost as strongly non-adaptationist, with the obvious exception of the extreme neo-Darwinians such as Poulton, and of course later Fisher and Ford. Evolutionists today have mostly forgotten that as recently as the early 1940s, most evolutionists believed, along with Darwin, that conspicuous polymorphisms were non-adaptive. Any evolutionist today knows that Mayr considers the vast majority of conspicuous polymorphisms to be adaptations shaped by natural selection. Yet, consider what Mayr says in his *Systematics and the Origin of Species* (1942),

Neutral polymorphism is due to the action of alleles "approximately neutral as regards survival value". Ford (following Fisher) believes that this kind of polymorphism is relatively rare, because "the balance of advantage between a gene and its allelomorph must be extraordinarily exact in order to be effectively neutral". This reasoning may be correct in all the cases in which one of the alternative features has a definite survival value or at least is genetically linked with one. There is, however, considerable indirect evidence that most of the characters that are involved in polymorphism are completely neutral, as far as survival value is concerned. There is, for example, no reason to believe that the presence or absence of a band on a snail shell would be a noticeable selective advantage or disadvantage. Among the many species of birds which occur in several clear-cut colour phases (Stresemann 1926 and later papers), there is, with one or two exceptions, no evidence for selective mating or any other advantage of any of the phases.

Even more convincing proof for the selective neutrality of the alternating characters is evidenced by the constancy of the proportions of the different variants in one population. The most striking case is that of the snails *Cepaea nemoralis* and *C. hortensis*, in which Diver (1929) found that the proportions of the various forms from

Pleistocene deposits agree closely with those in colonies living today (Mayr 1942, p. 75).

Fisher and Ford strongly disliked this view of conspicuous polymorphisms in the 1930s and 1940s, and thought that Sewall Wright was the theoretician who had provided the modern justification, namely random drift, for such an interpretation. To combat Wright and this view, Fisher and Ford began their research on *Panaxia dominula* and Ford began the ambitious research programme that later provided the evidential basis for his monumental *Ecological Genetics* (1964 and later editions).

The details of the debate over *Panaxia dominula*, fascinating though they are, are too complicated for inclusion here (for a full account see Provine 1986, Chapter 12). The crux of the debate, however, is clear enough. Fisher and Ford carefully followed one population of the day-flying moth *Panaxia dominula* for the years 1941—1946, using the marking, release, and recapture method to generate data from which gene frequencies and effective population sizes could be estimated. They studied an easily observed polymorphism controlled by simple Mendelian inheritance, with the added convenience that all three Mendelian classes could be distinguished by sight. They found that the yearly fluctuations in gene frequency were too large to be caused by random genetic drift in a population of the size they had measured. They concluded, by elimination, that the observed changed in gene frequency must have resulted from fluctuations in natural selection. They concluded,

Thus our analysis, the first in which the relative parts played by random survival and selection in a wild population can be tested, does not support the view that chance fluctuations in gene ratios, such as may occur in very small isolated populations, can be of any significance in evolution (Fisher and Ford 1947, p. 173).

The denial by Fisher and Ford of *any* significance for random genetic drift in evolution on the basis of just one experiment spurred Wright to answer their challenge. Wright's defence of random drift took two very different forms, one of them new and characteristic of his future attitude toward the evolution of conspicuous polymorphisms. First, he challenged the adequacy of their data to support the conclusion that random drift could not be the cause of the observed fluctuations in gene frequency. Second, he used the very different argument that, even if the observed fluctuations were caused by fluctuations in natural selection, this held only for a single locus with two alleles in a case of conspicuous polymorphism and it did not follow that all genes varied in frequency for that same reason. "The situation is similar", Wright argued, "except for the element of intent, to one that is familiar to livestock breeders. With very intensive selection for particular characters, others must be allowed to vary at random if numbers are to be maintained" (Wright 1948, p. 285).

The debate over *Panaxia dominula* intensified, with angry exchanges on both sides (Fisher and Ford 1950; Wright 1951). More than anything else, the debate crystallized the differences between the Fisherian and Wrightian ways of thinking about evolution in nature. Each side was now even more motivated to produce supporting field research.

The effect upon Wright was to make him rethink the whole issue of conspicuous polymorphisms. Even before he became aware of the results of Cain and Sheppard, strongly indicating that conspicuous polymorphisms in colour and banding of the shell in *Cepaea nemoralis* were subject to strong selection pressures, Wright had already sent a letter to Cain arguing that conspicuous polymorphisms in general should be expected to have evolved under and remain under strong selective forces. Instead, he argued, conspicuous polymorphisms were a very small proportion of the genome where random drift would play almost no role; however, with the rest of the genome, random drift might be an important factor (Wright to Cain, 14 November 1950). Also, by the late 1940s, Wright had dropped his earlier view that random drift could cause non-adaptive differences between species. Random drift was still crucially important, generating novel interaction systems at the level of the local semi-isolated population. However, the action of natural selection, he argued, would cause such differences as might be observed even at the subspecies level to be adaptive. Thus, Wright fits well into the shift towards a more selectionist attitude that Stephen Jay Gould aptly describes as the 'hardening of the synthesis' in the late 1940s and early 1950s (Gould 1983).

INFLUENCE OF THE FISHER−WRIGHT CONTROVERSY UPON EVOLUTIONARY BIOLOGY

In this section I will support my earlier assertion that the controversy between Fisher and Wright had a great influence upon modern evolutionary biology. I should say immediately, however, that by no means all of the influence of Wright and Fisher came from their disagreements. Together with Haldane, Hogben, Chetverikov, and other quantitative evolutionists, Fisher and Wright had a very important joint influence upon evolutionary biology (Provine 1978). All the mathematical populationists agreed upon a number of specific points, such as the immense power of selection to change gene frequencies in an intuitively surprising small number of generations, the relative insignificance of mutation pressure in relation to selection pressure under most conditions, or the theory of balanced polymorphisms that flowed from heterozygote advantage. They also agreed, for the most part, upon which variables were the really important ones, such as selection rates, effective population size, or population structure. Finally, the work of the population geneticists was a crucial element in the vast narrowing of the controversies over the mechanisms of evolution in nature.

The evolutionary synthesis of the 1930s and 1940s certainly did not remove all controversy about mechanisms of microevolution or speciation, but what it did do, and resoundingly, was to greatly narrow the range of controversies that existed before 1930. An evolutionist like Henry Fairfield Osborn was a very prestigious man before the synthesis, but who now talks about 'Aristogenesis' or about any of the host of other theories that were so common and taken seriously by one or another major school of thought before 1930? The mathematical population geneticists played a central role in this narrowing of the possible mechanisms of evolution, primarily by demonstrating quantitatively that some mechanisms were not as powerful as they seemed intuitively, and others were totally superfluous. Within this narrowed scope of the mechanisms of evolution in nature, the controversy between Fisher and Wright did have an important and specifiable impact.

General Influence of the Controversy

The general influence of the Fisher—Wright controversy during the period of the evolutionary synthesis is deeply related to the long-standing debate about adaptive and non-adaptive mechanisms of evolution in nature. This issue has consistently attracted much attention from evolutionary biologists, from Darwin's day to the present (for a detailed account of the ongoing controversy about adaptive versus non-adaptive mechanisms of evolution, see Provine 1985a). During the period of the synthesis, the debate often became focused as 'Wright's concept of random drift versus Fisher's concept of natural selection', although Wright himself was advocating his shifting balance theory rather than simply random drift as an alternative to selection. Most of the major books and papers on general evolutionary theory during the synthesis period reflected the tension between either random drift versus selection or shifting balance theory versus selection in large random breeding populations. The latter was more faithful to the ideas of Wright and Fisher, but the former was more prevalent.

The tension between Fisher and Wright may easily be seen in such works as (in chronological order) Ford's *Mendelism and Evolution* (1931), Dobzhansky's *Genetics and the Origin of Species* (1937), Huxley's *The New Systematics* (1940) and *Evolution: the Modern Synthesis* (1942), Mayr's *Systematics and the Origin of Species* (1942), Simpson's *Tempo and Mode in Evolution* (1944) and Stebbins' *Variation and Evolution in Plants* (1950), all of which were influential works during the synthesis period. Although rather few young evolutionary biologists during the 1930s and 1940s read the technical papers of Fisher and Wright, they could not help being familiar with the tension between their views of evolution from the more accessible literature. Thus, the tension between Fisher and Wright was interwoven into the very fabric of evolutionary theory during this period.

Field Research on Genetics and Evolution in natural Populations

Fisher and Wright agreed on the centrality of certain variables in the evolutionary process, among them effective population size, selection rates, and population structure, but they disagreed strongly on the relative sizes of many of the variables they agreed were important. It was precisely the agreement on the crucial variables combined with disagreement on the sizes of the variables that provided such a great stimulus to field research on the genetics of natural populations. Theoretical population genetics could not settle the questions about sizes of variables; that required field research.

Before the early 1930s, those who studied natural populations (such as F. B. Sumner with *Peromyscus*), had few handles to guide their studies. But the pertinent variables of microevolution were clear by the mid-1930s. If a field researcher could only determine, for example, effective population size, then this would constitute a base for distinguishing between a Wrightian and a Fisherian pattern of evolution in the organism. Or if observing the frequency of a gene from year to year or season to season could be used in combination with measures of effective population size to estimate the relative roles of selection and random drift. Dobzhansky, Ford, Huxley, Timoféeff-Ressovsky, and others sounded the clarion call to use the variables pinpointed by the mathematical population geneticists to guide studies of natural populations that could in turn be used to discriminate between the models of evolutionary change proposed by the same population geneticists.

The experimental work that should test these mathematical deductions is still in the future, and the data that are necessary for the determination of even the most important constants in this field are wholly lacking. Nonetheless, the results of the mathematical work are highly important, since they have helped to state clearly the problems that must be attacked experimentally if progress is to be made. . . . The manner of action of selection has been dealt with only theoretically, by means of mathematical analysis. The results of this theoretical work (Haldane, Fisher, Wright) are, however, invaluable as a guide for any future experimental attack on the problem (Dobzhansky 1937, p. 121, p. 176).

The work of Fisher, Haldane, and Wright is of the greatest importance, showing us the relative efficacy of various evolutionary factors under the different conditions possible within the populations. It does not, however, tell us anything about the real conditions in nature, or the actual empirical values of the coefficients of mutation, selection, or isolation. It is the task of the immediate future to discover the order of magnitude of these coefficients in free-living populations of different plants and animals; this should form the aim and content of an empirical population genetics (Timoféeff-Ressovsky 1940, p. 104).

Both Dobzhansky and Timoféeff-Ressovsky were hoping primarily to discriminate between the evolutionary schemes of Fisher and Wright.

The field researches that began as attempts to discriminate between the evolutionary theories of Fisher and Wright are among the most important of the evolutionary synthesis period. No detailed analysis is possible here (for that see Provine, 1985b, 1986), but the primary studies on the genetics of natural populations of which I am speaking are Dobzhansky's 'Genetics of Natural Populations' series (Wright collaborated on five of the first fifteen; Lewontin et al. 1981), the collaboration of Fisher and Ford on *Panaxia dominula* (Fisher and Ford 1947), of Cain and Sheppard on *Cepaea nemoralis* (1950), the work of Lamotte on *Cepaea nemoralis* in France (Lamotte 1951), and the ambitious, but abortive attempts of Buzzati-Traverso *et al.* to study the genetics of natural populations in Italy (1938). All of these except the last were ongoing projects, some continuing to the present day and still reflecting their origin in the controversy between Fisher and Wright.

One way to see graphically the great impact the tension between Fisher and Wright had upon the study of genetics and evolution in natural populations is to examine the first edition of Ford's *Ecological Genetics* (or the later editions). Even a cursory reading will reveal that most of the field research described in the book was begun with the controversy between Fisher and Wright explicitly in mind. The book is therefore in one sense a monument to the productive stimulus that the controversy provided.

CONCLUSIONS

Controversy in science frequently produces nothing but harsh words, hurt feelings, unpleasantries, and backbiting. Yet at other times, controversies manifesting many of these same symptoms have very positive and stimulating effects upon the whole field, and are central to later developments. The controversy between Fisher and Wright was one of these.

Indeed, the controversy between Fisher and Wright was in my opinion so central to modern evolutionary biology that it has become invisible to young people in the field. I always recommend that any aspiring graduate student in the field of evolutionary biology read Wright's four volume *Evolution and the Genetics of Populations* (in reverse order!), the second edition of Fisher's *Genetical Theory of Natural Selection*, and his papers on evolution from the *Collected Papers of R. A. Fisher*, (Bennett 1971–1974). Those who complete this exercise are invariably much impressed by how alive and central are many of the issues upon which Wright and Fisher disagreed to current evolutionary theory and research on natural populations.

ACKNOWLEDGEMENT

* This article has been published in *Oxford Surveys in Evolutionary Biology*, 1985, vol. 2, pp. 197–219, reprinted here by permission of Oxford University Press.

Division of Ecology and Systematics Cornell University

REFERENCES

Bennett, J. H. (ed.) (1971—1974), *Collected papers of R. A. Fisher*, The University of Adelaide, Adelaide.

Bennett, J. H. (ed.) (1983), *Natural Selection, Heredity, and Eugenics*, Clarendon Press, Oxford.

Buzzati-Traverso, A., Jucci, C., and Timoféeff-Ressovsky, N. W. (1938), 'Genetica di popolazioni', *Consiglio Nazionale Delle Ricerche, La Ricerva Scientifica, Series II* **1**: 3—30.

Cain, A. J. and Sheppard, P. M. (1950), 'Selection in the Polymorphic Land Snail *Cepaea nemoralis*(L.)', *Heredity* **4**: 275—94.

Darwin, C. R. (1859), *On the Origin of Species*, John Murray, London.

Diver, C. (1929), 'Fossil Records of Mendelian Units', *Nature* **124**: 183.

Dobzhansky, T. (1937) *Genetics and the Origin of Species*, Columbia University Press, New York.

Fisher, R. A. (1918), 'The correlation between Relatives on the Supposition of Mendelian Inheritance', *Trans. Roy. Soc. Edin.* **52**: 399—433.

Fisher, R. A. (1922), 'On the Dominance Ratio', *Proc. Roy. Soc. of Edinburgh* **42**: 321—41.

Fisher, R. A. (1927), 'On Some Objections to Mimicry Theory — Statistical and Genetic', *Trans. Roy. Entomol. Soc. Lond.* **75**: 269—78.

Fisher, R. A. (1928a), 'The Possible Modification of the Response of the Wild Type to Recurrent Mutation', *Am. Nat.* **62**: 115—26.

Fisher, R. A. (1928b), 'Two Further Notes on the Origin of Dominance', *Am. Nat.* **62**: 571—4.

Fisher, R. A. (1929), 'The Evolution of Dominance: A Reply to Professor Sewall Wright', *Am. Nat.* **63**: 553—6.

Fisher, R. A. (1930), *The Genetical Theory of Natural Selection*, Clarendon Press, Oxford, Second edition, 1958.

Fisher, R. A. (1934), 'Professor Wright on the Theory of Dominance', *Am. Nat.* **68**: 370—4.

Fisher, R. A., and Ford, E. B. (1947), 'The Spread of a Gene in Natural Conditions in a Colony of the Moth *Paraxia dominula*', *Heredity* **1**: 143—74.

Fisher, R. A., and Ford, E. B. (1950), 'The "Sewall Wright Effect" ', *Heredity* **4**: 117—9.

Fisher Box, J. (1978), *R. A. Fisher: The Life of a Scientist*, Wiley, New York.

Ford, E. B. (1931), *Mendelism and Evolution*, Methuen, London.

Ford, E. B. (1964), *Ecological Genetics*, Methuen, London.

Gould, S. J. (1983), 'The Hardening of the Modern Synthesis', in *Dimensions of Darwinism* (ed. M. Grene), pp. 71—93, Cambridge University Press, Cambridge.

Haldane, J. B. S. (1964), 'A Defense of Beanbag Genetics', *Persp. Mod. Biol. Med.* **7**: 343—60.

Huxley, J. S. (ed.) (1940), *The New Systematics*, Oxford University Press, Oxford.

Huxley, J. S. (ed.) (1942), *Evolution: the Modern Synthesis*, Allen and Unwin, London.

Jacot, A. P. (1932), 'The Status of the Species and the Genus', *American Naturalist* **66**: 346—64.

Lamotte, M. (1951), *Recherches sur la structure génétique des populations naturelles de Cepaea nemoralis L*, Supplement au *Bulletin Biologique de France et de Belgique*, No. 35.

Lewontin, R. C., Moore, J. A., Provine, W. B., and Wallace, B. (1981), *Dobzhansky's Genetics of Natural Populations*, Columbia University Press, New York.

Mayr, E. (1942), *Systematics and the Origin of Species*, Columbia University Press, New York.

Provine, W. B. (1978), 'The Role of Mathematical Population Geneticists in the Evolutionary Synthesis of the 1930s and 1940s', *Stud. Hist. Biol.* 2: 167—92.

Provine, W. B. (1985a), 'Adaptation and Mechanisms of Evolution after Darwin: A Study in Persistent Controversies', in *The Darwinian Heritage* (ed D. Kohn), pp. 825—66, Princeton, Princeton University Press.

Provine, W. B. (1985b), 'The Study of the Genetics of Natural Populations During the Evolutionary Synthesis of the 1930s and 1940s in *La vita e la sua storia* (ed. L. Bullini, M. Ferraguti, F. Mondella, and A. Oliverio), pp. 121—8. *Scientia*.

Provine, W. B. (1986), *Sewall Wright and Evolutionary Biology*, University of Chicago Press, Chicago.

Robson, G. C. (1928), *The Species Problem*, Oliver and Boyd, London.

Simpson, G. G. (1944), *Tempo and Mode in Evolution*, Columbia University Press, New York.

Stebbins, G. L. (1950), *Variation and Evolution in Plants*, Columbia University Press, New York.

Stresemann, E. (1926), 'Übersicht über die "Mutationsstudien" I—XXIV und ihre wichtigsten Ergebnisse', *J. Ornith.* 74: 377—385.

Timoféeff-Ressovsky, N. W. (1940), 'Mutations and Geographical Variation', in *The New Systematics* (ed. J. S. Huxley), pp. 73—136. Clarendon Press, Oxford.

Wright, S. (1921a), 'Systems of Mating. I. The Biometric Relation between Parent and Offspring', *Genetics* 6: 111—23.

Wright, S. (1921b), 'Systems of Mating. II. The Effects of Inbreeding on the Genetic Composition of a Population', *Genetics* 6: 124—43.

Wright, S. (1921c), 'Systems of Mating. III. Assortative Mating Based on somatic resemblance', *Genetics* 6: 144—61.

Wright, S. (1921d), 'Systems of Mating. IV. The Effects of Selection', *Genetics* 6: 162—6.

Wright, S. (1921e), 'Systems of Mating. V. General Considerations', *Genetics* 6: 168—78.

Wright, S. (1929a), 'Fisher's Theory of Dominance', *Am. Nat.* 63: 274—9.

Wright, S. (1929b), 'The Evolution of Dominance', *Am Nat.* 63: 556—61.

Wright, S. (1930), *The Genetical Theory of Natural Selection*, by R. A. Fisher (review). *J. Hered.* 21: 349—56.

Wright, S. (1931), 'Evolution in Mendelian Populations', *Genetics* 6: 97—159.

Wright, S. (1932), 'The Roles of Mutation Inbreeding, Crossbreeding and Selection in Evolution', *Proc. 6th Int. Cong. Genetics*, 1: 356—66.

Wright, S. (1934a), 'Physiological and Evolutionary Theories of Dominance', *Am. Nat.* 68: 25—53.

Wright, S. (1939), 'The Distribution of Self-Sterility Alleles in Populations', *Genetics* 24: 538—52.

Wright, S. (1948), 'On the Roles of Directed and Random Changes in Gene Frequency in the Genetics of Populations', *Evolution* 2: 279—94.

Wright, S. (1951), 'Fisher and Ford on the Sewall Wright Effect', *Am. Scient.* **39**: 452–8, 479.

Wright, S. (1967), 'The Foundations of Population Genetics', in *Heritage from Mendel* (ed. R. Alexander Brink), pp. 245–63. University of Wisconsin Press, Madison.

Wright, S. (1968), *Evolution and the Genetics of Populations. Vol. 1. Genetic and Biometric Foundations*, The University of Chicago Press, Chicago.

Wright, S. (1969), *Evolution and the Genetics of Populations. Vol. 2. The Theory of Gene Frequencies*, The University of Chicago Press, Chicago.

Wright, S. (1977), *Evolution and the Genetics of Populations. Vol. 3. Experimental Results and Evolutionary Deductions*, University of Chicago Press, Chicago.

Wright, S. (1978a), *Evoluation and the Genetics of Populations. Vol. 4. Variability within and among Natural Populations*, University of Chicago Press, Chicago.

Wright, S. (1978b), 'The Relation of Livestock Breeding to Theories of Evolution', *J. Anim. Sci.* **46**: 1192–200.

Wright, S. (1982), 'The Shifting Balance Theory and Macroevolution', *Ann. Rev. Genetics* **16**: 1–19.

M. J. S. HODGE

BIOLOGY AND PHILOSOPHY (INCLUDING IDEOLOGY): A STUDY OF FISHER AND WRIGHT

1. INTRODUCTION: HISTORIOGRAPHY, BIOLOGY AND PHILOSOPHY

The task undertaken in this paper will make better sense if I relate it to two familiar rationales for doing any work in the history of science.[1] They are rationales, indeed, that are matched in doing the history of many things besides science. The first is that if there is no collective quest for critical history, then myths and legends flourish, so that we have to settle for Rusk's version of Vietnam or Thatcher's invocations of the Victorian era. The second is that the future is unavailable and the present transient, so that the past is the only long run accessible to us. If we wish to understand how something — science, the economy or whatever — goes over the long haul, it is to the past that we must turn.[2]

Now, there are currently under active discussion a number of proposals about the long run of evolutionary theorising in biology, Three, in particular, will concern us here.

First, several writers, including Garland Allen and John Greene, have urged that the original Darwinism of Darwin's supporters should be read as mechanistic and materialist. The question has been raised, therefore, as to how subsequent evolutionary theorists stand in regard to that mechanistic materialism. Allen suggests that some have progressed beyond it to holistic materialism, others to dialectical materialism. Not surprisingly, however, Allen's and Greene's shared assumption has been challenged. Most recently, it has been challenged by Robert Richards in his comprehensive history of evolutionary theories of mind and behavior. Anyone discussing such figures as R. A. Fisher (1890–1962) and Sewall Wright (1889–1988) has to decide, then, whether they conform to the Allen historiography or, rather, to Richards's alternative.[3]

Second, Peter Bowler has argued that there was an evolution revolution in the nineteenth century, but that it was a non-Darwinian revolution, in that there was no widespread acceptance of Darwin's own views about the ramifying, directionless course and opportunistic,

231

Sahotra Sarkar (ed.), The Founders of Evolutionary Genetics, 231–293.
© 1992 *Kluwer Academic Publishers. Printed in the Netherlands.*

selectional causation of evolution. Later, Bowler holds, there was, in the twentieth century, a Mendelian revolution that moved evolutionary theory away from nineteenth-century, non-Darwinian views; especially away from non-Darwinian developmentalism, the interpretation, in other words, of evolution as a process analogous to individual development. So, on Bowler's account, it was this Mendelian revolution that prompted a delayed Darwinian revolution, by allowing something like Darwin's own non-developmentalist position to become widely adopted for the first time. We are invited, accordingly, to ask how far any mid-twentieth-century evolutionary theorists conform to Bowler's thesis.[4]

Third, there are now several volumes devoted to what has been called the Probabilistic Revolution, that shift in thought whereby chance and chances, matters stochastic and statistical, rose from marginal insignificance in the early seventeenth century to pervade everything from science and politics to sport and commerce in our own time. Evolutionary theorising has often involved probabilistic thinking, so that any study of particular evolutionary theorists offers an occasion to confirm or correct various views now current as to how probabilistic thinking in the sciences has developed over the long run.[5]

Fisher and Wright turn out to be highly instructive figures when related to such general historiographical issues as are raised by these three proposals. Most especially, we can be enlightened if we approach them, as biographers have: that is, with no predeliction for dividing their work from their lives, or for dividing their scientific work from their philosophical and even their ideological sympathies and commitments.[6] We can decline such predelictions and take a more comprehensive and integrated view, moreover, without making any presumption of explanatory priority in favor of one element or another in the biographical picture; without, say, presupposing that scientific positions are always dictated by prior religious beliefs, or that methodological reflections are always *post hoc* rationalisations of prior scientific convictions.

The desirability of not making any such presumption in the present instance is apparent once we consider what understanding of the term philosophy is appropriate here. In one sense, a scientist's philosophy is what is manifested in his or her practice of construing in a distinctive way very general questions about some scientific subject. Fisher is well known, for example, for comparing and contrasting his Fundamental Theorem of Natural Selection with the Second Law of Thermodynamics. To examine the presuppositions Fisher was making in such comparisons and contrasts is to examine his philosophy. In another sense, a scientist's philosophy is manifested in any position he or she takes on such issues as mind in relation to body or determinism and free will. To examine Fisher and Wright's views on these traditional issues is to examine their philosophies. Finally, there is a third sense of philosophy, one that includes ideology. Anyone's philosophy, whether scientist or otherwise, is being examined, in this sense, when we look to see how he or she stands in relation to liberalism or to socialism, the exemplary ideologies or social philosophies. It is a measure of how pervasive are philosophical themes, in the thinking of Fisher and Wright, that for our purposes here we need to consider all three senses of the term philosophy.

Professional scientists in the twentieth century do not usually develop and publish explicit philosophical views. To that extent Fisher and Wright are not typical, however, we should not let that point mislead us into thinking that their philosophical reflections were seen by them as peripheral excursions confined to moments when they were off duty. Indeed, we have to avoid begging any questions as to how their philosophy was related to their science. The relationship was, indeed, very different in each case; and that difference is something to be paid close attention, rather than given some characterisation grounded, *a priori*, in general preconceptions as to how professional science and amateur philosophy are likely to be associated with one another.

Let it suffice here to emphasise that Fisher and Wright were both serious amateurs when it came to philosophy. Wright prepared for his biographer Provine a table of interests and achievements for himself and others in the Wright family. As with mathematics and theoretical biology, Wright gave himself four stars underlined for philosophy. His autobiographical remarks in his philosophical writings make it clear that the reading decisive for his views on metaphysical issues was mostly done in the period 1912—14, and so before he had completed his doctoral studies at Harvard. Only much later, however, did he give those views a public airing. To the surprise of his audience, he devoted his Presidential Address to the American Society of Naturalists, in 1952, to a discourse that ran all the way from the hierarchy of organisms — including cells and species as organisms — to a dual aspect panpsychist analysis of mind, will and freedom.[7] Likewise, Fisher's biography, his correspondence and his reviews, his articles and his lectures, show that he was actively reading and reflecting on philosophical subjects from his student days on. By the 1930's he was making a full debut as a philosophical writer: in 1932 in a Herbert Spencer Lecture on the *Social Selection of Human Fertility* and in 1934 in an article, on 'Indeterminism and Natural Selection', for the first volume of *Philosophy of Science*. These publications had a sequel in 1950 in his Eddington Memorial Lecture: *Creative Aspects of Natural Law*.[8] Neither Fisher nor Wright felt any reticence or embarrassment regarding philosophy. For neither of them, then, did the designation of some view as philosophical carry any derogatory connotations. Quite the reverse. Reviewing Wright's first, full-length exposition of his evolutionary theory — the 1931 paper, 'Evolution in Mendelian Populations' — Fisher explained that, apart from "the scientific conclusions" established independently by workers in several countries, Wright "makes some philosophical observations on the nature of the evolutionary process, which are of great interest, although necessarily more personal and subjective". Fisher then gave a sympathetic but critical account of Wright's conclusions about the respective roles in evolution of factors making for heterogeneity and factors making for homogeneity: the conclusions later known as Wright's shifting balance theory of evolution.[9] Fisher's categorising of these views as philosophical implied no castigation. The two men were still on very friendly terms at this time. When engaged in discussing, even without agreement, the conditions most favorable for adaptive evolution, both saw themselves as supplementing and complementing secure scientific results with more uncertain conclusions that were properly called philosophical but none the worse for that.

2. A DECADE OF COLLABORATION AND AMICABLE DISAGREEMENTS: 1924—34

A study of Fisher's and Wright's biology and philosophy does best to begin with Fisher. Once we have a sense of Fisher's views, we can ask how and why Wright agreed and disagreed with him as he did. To proceed thus is historically and biographically appropriate, because Fisher was one step ahead of Wright at a crucial moment; so that Wright was making up his mind about Fisher's thinking rather than the other way round.

The crucial moment came in 1922. For it was then that Fisher published his paper 'On the Dominance Ratio'. Despite its misleadingly narrow title, this paper gave, in fact, a general treatment of population genetics. It was, moreover, the first paper to analyse the genetics of populations in the way that was quickly to become standard. For Fisher here considers such causes of change as selection and random sampling error insofar as they influence the statistical distribution of gene frequencies. So, a population is treated as a collection of genes, with each gene having a certain frequency because present in a certain proportion of individuals; and it is enquired what the statistical distribution of those gene frequencies is. Thus, if that distribution is a normal distribution, so called, as represented by the familiar bell curve, then many genes will be present in about half the individuals, while only a few will be present in either a great majority or a small minority. Such an analysis allows, therefore, for a quite general and abstract representation of the variability of a population.

Evolution, on such a representation, can be analysed as change in the distribution of gene frequencies. For, under Mendelian assumptions, the distribution is stable in a large population with random mating and no mutation, no selection and no migration. Accordingly, in this paper, Fisher did what no one had done before. He asked how such factors as dominance relations, mutation, selection and random extinction of genes in finite populations would affect the distribution; and he derived expressions for the effects of various mutation rates or selection intensities and so on. He also hinted at a conviction that he would never give up: namely, that adaptive evolution is most effectively produced in a large, randomly breeding population subject to sustained natural selection of very small heritable differences.

The strength of this conviction was already apparent in Fisher's review, the year before, of A. L. and A. C. Hagedoorn's book *The Relative Value of the Processes Causing Evolution*. However rare mutations may be, they can still be an "important factor in maintaining the variability of species", Fisher insisted, because "their frequency of occurrence will increase proportionately to the number of individuals in the species". A "very small rate of mutation" will suffice in "a large population" both to "continually supply new forms for the action of natural selection" and to "counterbalance" very easily the loss of variability through random extinctions of genes.[11]

The importance of large population size, for Fisher's understanding of the respective roles of random and selectional influences on the fate of mutant genes, is apparent also in a short, popular piece written at this same time. In 'Darwinian Evolution of Mutations', Fisher explained how he saw the new Mendelism allowing for a vindication of Darwin's selectionism. For this vindication, two novel findings are decisive, Fisher urged. First, the ordinary differences between parents and offspring are due not to new influences producing new characters, but to the segregation of existing Mendelian genes yielding new gene combinations. Second, new genes do arise but only very infrequently; however, heredity being particulate not blending, the maintenance of heritable variability is ensured by segregation, and so does not require a high rate of mutation. Any new gene arising by mutation — "in a single individual of population consisting of some thousands of millions" — will have its populational fate determined partly by chance, partly by selection. In an early phase, even if the gene is subject to favorable or unfavorable selection, chance will play the major role, because the initial frequency is so low. However, if the mutant gene survives this phase, and if it is advantageous on average, then selection will ensure that its frequency continues to rise, even though the advantage is very small. Hence, then, the benefit of sexual reproduction: mutation is "a leap in the dark", much more likely to fail adaptationally than succeed, especially if it has a large effect. The benefit of sex comes from maintaining the variability of a species with the minimum of mutations; with, that is, the "greatest stability of the reproductive processes". For consider any population "differing in a great many Mendelian factors, as all sexual populations are found in nature to do;" here "a single mutation may enable thousands of genetic combinations to be tested, and if any of these should happen to be very advantageous, it will by selection become the predominant type." Thus is "manifest" the benefit of "Mendelian inheritance of sexually reproductive organisms", most especially when "complex adaptations have to be made to a slowly changing environment," as Fisher argued.[12]

These three publications of 1921—1922, taken together, constitute an opening initiative by Fisher in his incipient campaign to establish a Darwinian view of evolution on the new foundations provided by his own prior synthesis of biometry and Mendelism. Readers of his earlier publications could have discerned his Darwinian loyalties, but only in 1922 could they have recognised the main strategies in the new campaign.

Two years after the dominance ratio paper appeared, Fisher and

Wright met and talked in the United States. Wright had not seen the paper, but Fisher sent him an offprint on returning to England. Wright then set about making his own analysis of the statistical distribution of gene frequencies. His mathematical techniques were very different from Fisher's. Integral and differential calculus had been Fisher's main resources. Wright used his own method of path coefficients that he had pioneered in previous papers on inbreeding, outbreeding and selection. However, despite these differences in the mode of mathematical attack, the two men converged on explicit agreements, often developed in correspondence, about the quantitative treatment of mutation rates, selection and so on as determining the statistical distribution of gene frequencies. By 1931 they were satisfied that the last mathematical discrepancies had been resolved.[13]

By this time Fisher had published his book of 1930, *The Genetical Theory of Natural Selection*, and Wright his long 1931 paper on 'Evolution in Mendelian Populations'. Now, Wright was composing a very full review of Fisher's book, while Fisher was writing a short review of Wright's paper. What the book, the paper and the two reviews, together with the correspondence, make plain is that despite that mathematical consensus on those quantitative topics, there was no agreement at all as to the conditions most favorable for adaptation and progress in evolution. For while Fisher was arguing, as he had ten years before, for a large randomly mating population subject to selection, Wright concluded that the optimal condition was a large species broken up into small partially-isolated local populations with a considerable degree of inbreeding within these populations, and low rates of inter-breeding among them. In his review of Fisher, Wright summarised Fisher's account of the transformation and multiplication of species in evolution, and declared Fisher's "conception of evolution" to be "pure Darwinian selection". Distancing himself from the "indefinitely large population" of Fisher's "scheme", Wright urged that, on his own scheme, there would be "a rapid differentiation of local strains, in itself non-adaptive, but permitting selective increase or decrease of the numbers in different strains and thus leading to relatively rapid adaptive advance of the species as a whole". Thus between "the primary gene mutations, gradually carrying each locus through an endless succession of allelomorphs, and the control of the major trends of evolution by natural selection", he would "interpolate a process of largely random differentiation of local strains".[14]

It is well known that this disagreement between Fisher and Wright
was to prove enormously influential and fruitful over the next half
century and more. What needs to be emphasised here is its independence from the mathematical consensus between the two men. The
disagreement is not traceable to differences in their mathematical
techniques. On the contrary, it existed despite the mathematical consensus they had reached by different technical routes. Nor can personality conflicts be invoked at this juncture. Later, starting in 1934, Fisher
and Wright fell out personally and irreversibly, so as to be incapable of
discussing any differences of opinion amicably. But, before that falling
out, there was a whole decade of collaboration and amicable disagreement. Fisher's biographers Box and Bennett and, even more definitively, Wright's biographer Provine have established the character and
content of the interaction over this decisive decade; and the account
given here draws throughout on Provine's writings. What our comparative and contrastive biographical task next requires of us is therefore
plain enough. We need to look at the biology and philosophy of both
men in the years before that decade. By doing that we can hope to
understand better the remarkable mixture of agreement and disagreement the emerged in those years.[15]

3. FISHER: SCIENTIFIC AND PHILOSOPHICAL THEMES
OVER HALF A CENTURY

It is now well established that the undergraduate Fisher (1909—12) had
already developed four clusters of interests — apart from those in his
degree subject, mathematics — that were to last him the rest of his life:
Mendelism, biometry, Darwinism and eugenics. It is also clear that
those commitments were being actively integrated even at this early age.
Nor are the main lines of the integration in doubt. Mendelism and
biometry were not to be opposed to one another, as they were still by
many in England, including the leading Mendelian, Bateson and the
leading biometrician, Pearson. Rather, as Fisher saw it, the Mendelian
account of hereditary factors, to be known in a few more years as
Mendelian genes, was to be reconciled with the biometric estimates of
the correlations among the quantitative characteristics of relatives,
whether parents and offspring, siblings or cousins. Moreover, the
principal premise in this reconciliation was already the one that would
be made in 1916, when Fisher wrote the paper of 1918, 'On the

Correlation of Relatives on the Supposition of Mendelian Inheritance':
the premise, that is, that the continuous quantitative variation in a
strongly heritable character, such as height appeared to be in some
human populations, is due to many Mendelian factors each of small
effect.

Given this reconciliation of Mendelism and biometry, the integration
of Mendelism and Darwinian selectionism could take, also, the form it
would take later: that is, with Mendelian factor differences providing
for the heritable variation that would be the material for selective
breeding, as practised by man on the farm or entailed by the struggle
for existence in nature.

For the young Fisher, this Mendelian selectionism offered, in turn, to
integrate evolution with eugenics. Taking the human species to be
originally a gradual, natural product of Mendelian heredity worked on
by Darwinian selection, he could ask how far selection among different
human societies may make for social progress in the future.[16]

Now, although this much about the undergraduate Fisher is plain, at
least in outline, we have to be wary of finding all his beliefs and attitudes, even of the early 1920's anticipated ten years before. To give just
three examples, a discriminating history of his thinking over the decade
after Cambridge would have to emphasise how his eugenic views
changed markedly as individual selection within societies joined selection among societies as a principal concern; it would have, too, to
emphasise how his Mendelism expanded from being merely a theory of
factors subject to linkage and segregation and so on, to being a theory
of genes that included, following Muller especially, new conclusions
about the nature and rates of gene mutations; and how his selectionism
was developed in contrasting gene frequency changes — and so gene
frequency distribution changes — due to selection, and those due
to such chance accidents of differential survival and reproduction
("Hagedoorn effect" Fisher would call it first) as would eventually, in
later years, be dubbed random genetic drift.

We also have to be wary of tracing Fisher's interests to a single
motivational source. Some years ago, I agreed with some others in
denying that Fisher was, in any strong sense, an evolutionary biologist;
recommending that we think of him, instead, as a Mendelian biometrician whose sole ultimate motivation was to establish a new Mendelian
eugenics. I am now convinced that this recommendation is inconsistent
with the full biographical picture, and that it is wrong in suggesting that

the two characterisations — evolutionary biologist and Mendelian eugenist — might be exclusive of one another. Fisher was both of these things, and he was not either of them to a lesser degree because he was the other. On the contrary, his commitment to the understanding of the causes of evolution and his commitment to the improvement of society — or at least to the prevention of deterioration in English society — were complementary and mutually reinforcing, although also sufficiently distinct if not separable that their sources require distinct, albeit integrated, biographical treatments.[17]

To see something of what such a treatment might involve, it will be appropriate to concentrate briefly on Fisher's selectionism. The depth and intensity of Fisher's preoccupation with selection is familiar to all who have read in, or even about, the man. However, it will be urged here that this depth and intensity were even more extreme than has yet been recognised. For a start, recall Fisher at high school choosing a set of Darwin's volumes as a prize award. This moment alerts us to Fisher's lifelong tendency to identify evolution as a topic with Darwin as an author, in that there will be for him no making up one's mind about anything concerning evolution without making up one's mind about Darwin's view of evolution. Fisher's passionate, perennial concern with evolution and Darwin constitute good grounds for insisting that he was an evolutionary biologist. The identification of evolution and Darwin entailed, for Fisher, another identification, because he always read Darwin as making one supreme contribution to evolution as a topic: namely, natural selection. So evolution, Darwin and natural selection turn out to be, for certain purposes, one topic. Now, Fisher did open the very Preface of his book *The Genetical Theory of Natural Selection* of 1930 by declaring evolution and natural selection to be distinct, and by proposing that natural selection be studied in its own right and independently from its explanatory deployment as a cause of evolution. However, in making natural selection an independent topic, Fisher did not make evolution one, too.

4. ADAPTATION IN A TWO-TENDENCY UNIVERSE

As dozens of Fisher's publications show, he was always of the view that evolution was primarily an adaptive process — a continuing adjustment of complex structures and actions to environmental change — so that the primary constraint on a theory of evolution was that it explain adaptation as a process and so as a product. Natural selection ("Darwinism") and the inheritance of acquired characters ("Lamarckism") were the only two theories to really offer to meet this constraint, Fisher liked to argue; but since Darwin's day Lamarckism had been shown to be inconsistent with what is known of variation and heredity. There can, then, be no acceptable thinking about evolution that is not primarily thinking about natural selection; a causal process, as Fisher recalled, that had been valued, and rightly, by the very first Darwinians as a "known cause", one that is evidenced independently of the facts it was invoked to explain.[18] Fisher's writing conforms consistently to this limitation, in that whenever evolution is the subject, the theory of natural selection — as the only acceptable theory of adaptation, and so of evolution — is what has to be discussed. Fisher accordingly once opened a discussion on the theory of natural selection by going so far as to say:

Theories of Evolution are of two kinds, those that, in Professor Watson's words, "are explanations primarily of adaptation and only secondarily of the origin of species", and those which fail to account for adaptation. To the first class belongs the theory of the inheritance of acquired adaptations called Lamarckism, and the theory of the natural selection of innate adaptations. For these two theories evolution *is* progressive adaptation and consists of nothing else. The production of differences recognisable by systematists is a secondary by-product, produced incidentally in the process of becoming better adapted.[19]

Now, we can not discern just when and how Fisher first reached this extreme adaptationist-selectionist position. The evidence suggests, however, that it was always there, in that it did not replace any earlier, different position; it was, rather, that as his thinking about evolution developed over the Harrow and Cambridge years, this outlook came to be more and more an entrenched foundation for any subsequent reflections. What we can be more sure of is that, from 1912—22, two presences in his life were to work upon Fisher so as to reinforce, even more strongly, this incipient foundation. Of the two, the first in time was statistical mechanics as represented by an eminent master of that branch of physics, James Jeans, Fisher's principal teacher at Cambridge for a postgraduation year (1912—13) devoted to statistical mechanics and quantum theory. The second was the Darwinian heritage and legacy itself, as represented by Leonard Darwin, a childless mentor who came, in effect, to adopt Fisher as the grandson of Charles Darwin that he never had himself, but whose place Fisher quickly came to take within a year or two of moving from Cambridge to London.

This conjunction — of statistical mechanics and Darwinian patronage — proves to be such a telling one for any effort to understand Fisher's entire biological and philosophical outlook, throughout his whole life, that it is as well to anticipate, here, something of the subsequent analysis, in order to convey in outline its larger significance. One strategy in doing this, in dramatic terms, is to say that Fisher's universe is going to be, ultimately, a two-tendency universe. There will be the entropic tendency which, if not countered, takes everything toward more probable, and less organised, states. This is the way down. Unique in its countering of such degenerative, deteriorational trends is the counter-entropic tendency entailed by selection as it takes anything subject to its workings toward less probable and more organised states. This is the way up.

The dramatisation can be made more personal, because the way down and the way up have each a principal interpreter; so that Fisher's natural science has ultimately a two-hero history: Ludwig Boltzmann (1844—1906), as leader of a triumvirate also including James Clerk Maxwell and Willard Gibbs, and Charles Darwin (1809—1882) senior member of a partnership with Wallace. For these are the nineteenth-century authors of the two great probabilistic insights that have, as Fisher saw it, set the decisive precedents for all subsequent thinking about the inanimate and animate creation, including man himself.

To talk of creation and of man as a creature, who is himself creative, is entirely in order here. Fisher never wavered in the Anglican, Christian faith he was raised in, and he was eventually to preach occasional sermons in his college's chapel at Cambridge during his years there as Professor of Genetics. His science always sought causes, and causation he would view, quite explicitly, as probabilistic and creative, indeed creative because not deterministic but indeterministic. His indeterminism was to be developed to integrate his Christianity, with its theory of creation and creativity, and his science, his Boltzmannian and Darwinian science. Likewise, in integrating his Christianity with his eugenics: the indeterminism allows for man to emulate God's own work by freely intervening in social life, in the institution of those eugenic measures that alone can ensure the permanency of English civilisation; for these measures — suitable family allowances, say — can offset the social decline otherwise arising from the diminished fertility that accompanies the greater contribution to society made by those gifted individuals who have risen to take their rightful place in the middle classes. Here, Fisher joins his scientific and political forces with Major Leonard Darwin, whose zeal and energy on behalf of the eugenics movement were often directed to the membership of the Eugenics Society whose President he was from 1911 to 1928. Gratefully and enthusiastically did Fisher, for his part, discharge his filial and grand-filial roles as eugenist and evolutionary biologist in the years culminating in the publication in 1930 of *The Genetical Theory of Natural Selection* with its dedication to the former President. What is more, for three more decades until his death in 1962, Fisher was always building upon, and never repudiating, the outlook on the universe and on science represented in his career by this early conjunction of James Jeans and Leonard Darwin.

5. A SUGGESTED READING OF FISHER'S THEORY OF NATURAL SELECTION

To confirm this two-tendency and two-hero reading of Fisher's world and Fisher's thought, it proves useful to do three things, before leaving Fisher for Wright. The first is to offer some suggestions as to what may be implicit, although not explicit, in Fisher's understanding of evolution by natural selection.[20] Here, necessarily one cannot document directly the evidence for the interpretations offered, but one can attempt to show how those interpretations for various topics cohere into a plausible overall view of Fisher's position. The second is to look at moments when Fisher brings into explicit conjunction his thoughts about statistical mechanics — and related subjects such as thermodynamics and the kinetic theory of gases — and his scientific analyses of evolution by natural selection. Finally, the third is to follow Fisher himself in his overtly philosophical reflections, where, again, that conjunction is often brought to bear on his Christianity and his eugenics.

Before moving to the first of these tasks, there is a point to be made about all three. Throughout, the discussion will be abstract and general, in that rather little contact will be made with the precise details of what Fisher had to say about, say, mutation rates in *Drosophila* flies or mimetic resemblances in *Papilio* butterflies or the measurement of the average excess and average effect for any gene substitution. This lack of contact is in no way meant to suggest that such detailed matters are irrelevant. On the contrary, I would stress that one virtue in an abstract and general discussion is that it may allow for light to be thrown on such specific matters. Fortunately, Fisher's treatments of several such relevant matters have received extensive attention and clarification in a number of recent publications.[21] The present discussion is, therefore, designed to complement those publications, by being adapted throughout to what they have already done.

Turning now to the first task, it requires us to appreciate what may have been implicit in Fisher's privileging of natural selection as the sole counter-entropic agency at work in the natural world. One implication would have been that the optimal conditions for natural selection — and therefore, for Fisher, for evolution — would be the conditions wherein other, entropic factors are least frustrating of the work of natural selection. This consideration most concerns mutation and random genetic drift. For these are both disordering processes. Mutation is a disordering failure of exact replication in the genes of an individual. Random drift is a disordering failure in a population to exactly replicate its genetic structure.

Mutation is, obviously, a necessary condition, in the very long run, for the continued work of natural selection, because it is only by gene mutations that new genes, rather than new permutations of old genes, can arise. However, thanks to inheritance being particulate, not blending, very low rates of mutation can suffice to provide adequate material for selection, and the larger the population the lower the rate that suffices. Moreover, just as low rates are least disordering so are small effects; so, by the early 1920's, Fisher is emphasising that empirical findings about mutations make it reasonable to assume that mutations are low in their rates, small in their effects, very rarely advantageous and mostly initially masked in their effects because initially recessive. For their being all these things allows natural selection to be vindicated as the only cause that can determine the direction of evolution.

Random drift is an inevitable concomitant of selection, in that any population that is subject to selection is subject to drift. But the larger it is the less subject will it be to drift without being less subject to selection. Drift may be important in determining what mutations survive beyond initial rarity. But, beyond that, drift has no contribution to make to the adaptive adjustment of the whole of a large population to gradual environmental change.

So far what has been said concerning Fisher's thinking about genes and gene mutations may seem overly simple. But that is precisely what is to be grasped: namely, that Fisher prided himself as having seen that in its essentials this topic was a simple one, for the purposes of the theory of natural selection. For, after all, consider how few are the assumptions he feels it necessary to establish in Chapter One of the *Genetical Theory* — the chapter on 'The Nature of Inheritance' — before proceeding to the second chapter on 'The Fundamental Theorem of Natural Selection'. His strategy in that first chapter is, primarily, to introduce only the most basic assumptions about inheritance that were not available to Darwin and that can, now that they have been secured by recent genetical research, vindicate Darwin's theory, natural selection. The strategy, thereafter, is, first, to reach the most comprehensive statement possible about natural selection — the Fundamental Theorem — without involving mutation at all, as a special topic; and, then, second, to introduce mutation in three chapters, one on dominance as something that some mutations eventually acquire thanks to selection, and two on variation as determined by mutation and selection together. A further chapter on sexual reproduction and sexual selection can complete the quite general treatment of natural selection, before it is applied to mimicry and finally to man and society.[22]

The workings of natural selection can, then, be exhibited, throughout, as necessary and sufficient for all adaptive progress, right up to the moment, that is, when the conditions for a permanent human civilisation are confronted. For while natural selection can be shown, in Fisher's view, to ensure the progress from barbarous social life to civilised society, its continued action, if not actively countered by eugenic measures, will lead to an inevitable deterioration in any civilised society, thanks to the heritability of fertility and the social promotion of the less fertile.

Now, it is here that the earlier adaptationist stance has to be transcended. For, in order to understand the causes, consequences and conditions for countering this social promotion of the less fertile, Fisher has to engage the economics of social promotion and so of property, wealth and class in a society such as his own; and he has no way to represent these economic features of a civilised society as constituting an environment for the society to which it can be seen as adapting itself. Nevertheless, these features serve as a constraint on the eugenist, because Fisher takes the conservative liberal view that the best way forward is not to move to a different economic system, but to reverse the correlation between social promotion and infertility by introducing economic inducements, such as tax measures, that would provide sufficient incentives, in the existing economic system, for those individuals, whose heritable characteristics have enabled their social promotion, to raise more children and so keep up the populational frequency of the genes responsible for those characteristics. So, in this context, certain distinctive elements in Fisher's selectionism are plainly present: in particular, selection is worked through some influence exercised throughout a population, in a mass selection, therefore; however, selection is not acting counterentropically to produce and maintain highly improbable, complex adaptations to changing environmental circumstances.

Is Fisher's Mendelian selectionist eugenics for serving the permanence of civilisation in society a difficulty, then, for the reading given here of his adaptationist-selectionism? Surely not; for Fisher himself dwells explicitly on the reasons for a distinctive treatment of selection in civilised rather than barbarous societies. For one thing, differential fertility takes over from differential survival as decisive for the direction of change, once the shift is made out of barbarity into civilisation. This contrast entails, in turn, the contrast concerning adaptation as an adjustment to environmental circumstances; for whereas adaptation is always for survival and reproduction in, and so to, some particular circumstances, fertility as such is not interpreted as an accommodation to anything particular and circumstantial.

Such discontinuities, between the eugenics for civilised society and the genetical theory of natural selection for plant, animal and pre-civilised human life, suggest, therefore, two points about the interpretation of Fisher's biology and his eugenics. The first is that his evolutionary biology can not be reduced to any straightforward projection onto all nature of his eugenics. The second is that his eugenics had its own motivational sources in what is called here his conservative liberalism, with its meritocratic, possessive, individualistic assertion of middle class values within the existing economic organisation of society. An interpretative quest for the underlying unities in Fisher's thought must, therefore, look beyond his hereditarian, Mendelian selectionism as an element obviously common to his eugenics and his evolutionary biology. One is in a better position to discern what else has to be brought into such an interpretative quest, once a survey has been made of the comparisons and contrasts Fisher makes between his physics and his biology.

6. STATISTICAL MECHANICS, THERMODYNAMICS AND THE THEORY OF GASES, AND GENETICS AND THE THEORY OF SELECTION

The comparisons and contrasts that must concern us here are more diverse and complex than one might think at first. For a start, we have to notice when Fisher is invoking the precedent of statistical mechanics and when, rather, thermodynamics or, again, the kinetic theory of gases or, more generally, what he calls the kinetic theory of matter. Furthermore, he has various reasons for bringing these different precedents into his biological expositions; so that there is no way to reduce his invocation of them to a single formulation comprehending all cases.

One very early comparison appears in his undergraduate lecture, given to the Cambridge University Eugenics Society, on "Heredity". Biometrical work can yield beautifully certain results, he says. Dealing only in observations, biometricians avoid "all the difficulties of abstract theories"; moreover, although their observations are probably full of small errors, they seem able "to squeeze the truth out of the most inferior data". In every case the probable error can be calculated and, while possible error is unlimited, the probability of large errors is demonstrably very small. Fisher was "recently impressed" with this powerful feature of the theory of probabilities. Put a kettle on the fire and it will probably boil; not certainly, for it may freeze. The odds against this last outcome are very large, but it "remains a possibility, or so my 'theory of gases' tells me".[23]

Here, then, we have the young Fisher impressed with a commonplace analogy between probability in the theory of observational errors and in gas theory. A few years later, less commonplace arguments are appearing. In 1915, he and his co-author C. S. Stock, writing for the *Eugenics Review*, are castigating those who have opposed Darwinism to Mendelism, overlooking that whereas Darwinism is concerned with evolution, Mendelism concerns heredity rather than evolution as such. The mistake has led to serious misunderstandings regarding eugenics, and even ill-founded American legislation.

Raising as so often the issue of generality, Fisher insists that Mendelism is unlikely ever to "cover even the field of heredity"; while, by contrast, Darwin's theory does cover the field, in that it "explains and co-ordinates new facts". Mistakes, what is more, are more likely to arise from new specialised work such as the Mendelians pursue, than from a well-established theory such as Darwin's. Eugenists should hold to general principles; with Mendelism they are vulnerable to objections; on Darwin's ground they are not. The importance to eugenists of the broad principles of the *Origin* is paramount. Darwin's teaching, even if it were all that were available, says Fisher, "would not only allow but compel us to formulate eugenic concepts and proposals". The nature of this compulsion is clarified by two invocations of the kinetic theory of gases.[24]

The first is introduced to considering selection itself. Selection is the main cause of changes in the constitution of a mixed population. The existing and possible agencies of selection now, as always, provide the most fruitful field of eugenic research. "These agencies acting at large amidst a multitude of random causes", any one being the predominant influence on some particular individual, "nevertheless determine the progress or decadence as a whole". Likewise, then, in the "kinetic theory of gases", where the "several molecules are conceived to move freely in all directions with greatly varying velocities" but with a "statistical result that is a perfectly definite measurable pressure". The common feature here, then, shared by selection theory and gas theory is the reliability and predictability of the outcome when the individuals are numerous and the causes acting upon them independent. Allied to this lesson is the second. "Controversy may rage round the nature and properties of the atom yet our knowledge of general principles enables us to calculate gas pressures with accuracy". Here, we are "independent of particular knowledge about separate atoms, as in eugenics we are independent of particular knowledge about individuals". Of course such knowledge is important, interesting and useful, but it is, Fisher declares, "unnecessary alike for a general theory of gases and for a general theory of eugenics".[25]

This drive for generality in conforming to the precedents set by the theory of gases, is most explicit in Fisher's explanation, in his 1922 dominance ratio paper, of what he saw as his achievement in his 1918 paper on the correlation of relatives. That earlier paper attempted, he explained, "an examination of the statistical effects in a mixed population" of a "large number" of Mendelian factors. Now, apart from a continuing belief that the biometricians' results discredited Mendelian inheritance, there was another widespread misunderstanding regarding any such attempt. For it was generally believed that "the variety of the assumptions to be made about the individual factors" — assumptions about the dominance of particular factors, about the size of their effects, about their proportion in the population, about dimorphism and polymorphism, and about linkage, and also assumptions about "more general possibilities" such as homogamy, selection and environmental effects — made "it possible to reproduce any statistical resultant by a suitable specification of the population". It was important, therefore, Fisher recalled, "to prove that when the factors are sufficiently numerous, the most general assumptions as to their individual peculiarities lead to the same statistical results".[26] The stage is thereby set for the analogy with the theory of gases:

Although innumerable constants enter into the analysis, the constants necessary to specify the statistical aggregate are relatively few. The total variance of the population in any feature is made up of the elements of variance contributed by the individual factors, increased in calculable proportion by the effects of homogamy in associating together allelomorphs of like effect. The degree of this association, together with the quantity which we termed the Dominance Ratio, enter into the calculation of the correlation coefficients between husband and wife and. between blood relations. Special causes, such as epistacy, may produce departures, which may in general be expected to be very small from the general simplicity of the results; the whole investigation may be compared to the analytical treatment of the Theory of Gases, in which it is possible to make the most varied assumptions as to the accidental circumstances, and even the essential nature of the individual molecules, and yet to develop the general laws as to the behavior of gases, leaving but a few fundamental constants to be determined by experiment.[27]

From what we know of James Jeans's work on the theory of gases, he would have made sure that Fisher, as his pupil, acquired a keen sense of the latitude permissible, regarding assumptions about the circumstances and the properties of gas molecules, when one is carrying out these kinds of derivations. Fisher's understanding of this analogy with his 1918 derivation of the correlations of relatives may well trace, therefore, rather directly to what he had learned in that postgraduation year at Cambridge.

The 1922 paper contains in its own summary a further analogy: "the distribution of the frequency ratio" for different hereditary factors is — in the absence of selection and random survival effects and so on — a stable one like that of "velocities in the Theory of Gases". Years later, Fisher was more explicit about the change of viewpoint that this analogy made possible. The frequencies of different genotypes define the gene ratios for the population, so "it is often convenient to consider a natural population not so much as an aggregate of living individuals as an aggregate of gene ratios; such a change of viewpoint" being like that familiar in gas theory where specifying the "population of velocities" is often more useful than specifying "a population of particles".[28]

7. THE ANALOGIES AND DISANALOGIES IN THE GENETICAL THEORY

The analogy between the two stable distributions was duly developed further by Fisher, so that by 1930 in the *Genetical Theory* he was using it in contrasting the different implications of the blending theory of inheritance and of the particulate theory. The two theories give quite different interpretations of the fact that brothers and sisters, whose parentage and so entire ancestry is identical, may differ greatly in their hereditary constitutions. A blending theorist sees here evidence of new and frequent mutations associated — witness the greater resemblance of identical twins — with temporary conditions at conception and gestation. "On the particulate theory it is a necessary consequence of the fact that for every factor a considerable fraction, not often much less than one half of the population, will be heterozygotes, any two offspring of which will be equally likely to receive unlike as like genes from their parents". Given "the close analogy between the statistical concept of variance and the physical concept of energy", the heterozygote may be thought of "as possessing variance in a potential or latent form", so that rather than being lost on the mating of homozygous genotypes "it is merely stored" in a form allowing it to reappear later.[29]

A population mated at random immediately establishes the condition of statistical equilibrium between the latent and the apparent form of variance. The particulate theory of inheritance resembles the kinetic theory of gases with its perfectly elastic collisions, whereas the blending theory resembles a theory of gases with inelastic collisions, and in which some outside agency is required to be continually at work to keep the particles astir.[30]

We are now, finally, prepared to appreciate the significance for Fisher of the analogy he is going to draw between his Fundamental Theorem of Natural Selection and the Second Law of Thermodynamics. For the Fundamental Theorem relates the rate of increase in fitness, that selection produces in a population, to the genetic variance in fitness that selection consumes in producing that effect, genetic variance that would otherwise be conserved; just as the Second Law concerns the progressive loss of energy available for work.

Fisher's own main statement of his theorem in 1930 reads thus: "*The rate of increase in fitness of an organism at any time is equal to its genetic variance in fitness at that time*", with the word "species" used instead of "organism" in a recapitulation at the end of the chapter.[31] Now, much has been done since Fisher to clarify how he or anyone else may have derived this law from his or other assumptions. What concerns us, here, however, are not the possible derivations, but Fisher's contrasts and comparisons between his theorem and the second law of thermodynamics. For this purpose, two glosses are valuable. The first has been offered recently by Price. Given how Fisher's own derivation proceeded, says Price: "What Fisher should have written is something like this: 'In any species at any time, the rate of change of fitness ascribable to natural selection is equal to the additive genetic variance in fitness at that time'". For, according to Price, what Fisher's theorem says is that "natural selection (in his restricted sense meaning only additive effects) at all times acts to increase the fitness of the species to live under the conditions that existed an instant earlier".[32]

A second gloss was offered at the time by Wright who suggested that the simplicity of Fisher's formulation must not mislead us. For, he holds, Fisher's special sense of "genetic variance" means that dominance relations, epistatic effects and the effects of mutation and migration are not comprehended by the term and nor therefore by the theorem.[33]

Fisher himself stressed that the theorem is "exact only for idealized populations" with no fortuitous fluctuations of genetic composition. And he was keen to give an estimate of the size of the effect of these fluctuations, that is "a standard error" for the expected "rate of increase in fitness". He satisfies himself that this error will be small because such fluctuations will be small compared to the average rate of progress. Once again, he has his gas analogy. The "regularity" in the average rate of progress is guaranteed for the same reason that a bubble of gas obeys the gas laws. A visible bubble may contain billions of molecules, more than most animal and plant populations contain individuals, but "the principle ensuring regularity in the same", in that even if there are large fluctuations from one generation of a population to the next, over many generations the deviations will be small. By stating the theorem in the form he has, and specifying the relation between progress in fitness

— as measured by m, Fisher's Malthusian parameter — and its standard error, the objection can be met that natural selection "depends on a succession of favorable chances". Natural selection only depends on this in the same sense that the very reliable income of a casino owner does. There is every difference between "a succession of favorable deviations from the laws of chance" and the "continuous and cumulative action of these laws"; and it is on such a continuous and cumulative action that "the principle of Natural Selection relies". Beyond discrediting that old objection to natural selection, putting the theorem in the form Fisher gives it also allows him to make his multiple comparisons and contrasts with the thermodynamics.[34] First come the comparisons:

It will be noticed that the fundamental theorem proved above bears some remarkable resemblances to the second law of thermodynamics. Both are properties of populations, or aggregates, true irrespective of the nature of the units which compose them; both are statistical laws; each requires the constant increase of a measurable quantity, in one case the entropy of a physical system and in the other the fitness, measured by m, of a biological population. As in the physical world we can conceive of theoretical systems in which the dissipative forces are wholly absent, and in which the entropy consequently remains constant, so we can conceive, though we need not expect to find, biological populations in which the genetic variance is absolutely zero, and in which fitness does not increase. Professor Eddington has recently remarked that "The law that entropy increases — the second law of thermodynamics — holds, I think, the supreme position among the laws of nature." It is not a little instructive that so similar a law should hold the supreme position among the biological sciences.[35]

Then — after a hint that the way down and the way up might one day be brought within some even more general synthesis — come the contrasts, including, last of all, the contrast of the very greatest significance:

While it is possible that both may ultimately be absorbed by some more general principle, for the present we should note that the laws as they stand present profound differences — (1) The systems considered in thermodynamics are permanent; species on the contrary are liable to extinction, although biological improvement must be expected to occur up to the end of their existence. (2) Fitness, although measured by a uniform method, is qualitatively different for every different organism, whereas entropy, like temperature, is taken to have the same meaning for all physical systems. (3) Fitness may be increased or decreased by changes in the environment, without reacting quantitatively upon that environment. (4) Entropy changes are exceptional in the physical world in being irreversible, while irreversible evolutionary changes form no exception among biological phenomena. Finally, (5) entropy changes lead to a progressive disorganisation of the physical world, at least from the human standpoint of the utilization of energy, while evolutionary changes are generally recognised as producing progressively higher organization in the organic world.[36]

There, then, is a climactic text that takes on its full significance when we associate the broad sweep of its claims with Fisher's entire life and thought. Within two years after this he had been explicit in the characterisations of Boltzmann and of Darwin that accompanied this entire conception of science. Of Boltzmann, Fisher said:

Perhaps the most dramatic development [in the theory of gases] was when Boltzmann restated the second law of thermodynamics, the central physical principle with which so many of the laws of physics are interlocked, in the form that physical changes take place only from the less probable to the more probable conditions, a form of statement which seemed to transmute probability from a subjective concept derivable from human ignorance to one of the central concepts of physical reality. More concretely, perhaps, we may say that the reliability of physical material was found to flow, not necessarily from the reliability of its ultimate components, but simply from the fact that these components are very numerous and largely independent.[37]

Of Darwin he wrote that it was "his chief contribution, not only to Biology but to the whole of natural science" to have identified a process whereby "contingencies a priori improbable, are given, in the process of time, an increasing probability, until it is their non-occurrence which becomes highly improbable".[38]

8. THE BIOLOGY AND PHILOSOPHY OF A SCIENTISTIC ROMANTIC

One virtue of the two-tendency and two-hero reading of Fisher's universe and of Fisher's science is that it saves us from certain simplifying fallacies that we might otherwise commit. It would be tempting, for instance, to see Fisher's work in evolutionary biology as essentially the work of an alien invader, as the work, that is, of a man trained in mathematics and physics who would impose the methods, concepts and presuppositions of hard, exact science on a subject not naturally amenable to such a treatment. To be saved from this fallacious view of Fisher as evolutionary biologist, one needs only to see how characteristic of his entire outlook is his writing on such traditional Darwinian natural history topics as bird courtship or insect mimicry. For the obvious generalisation from Fisher's preoccupation with that kind of natural history topic is correct. He was not only in descent from Darwin, he rightly saw himself in descent from a Darwin who had had among his decisive ancestors the naturalists of truly olden days, indeed the parson naturalists and natural theologians of those olden days.[39]

Another virtue of this reading of Fisher is that we are saved from reading him as someone who sought to advance biology principally by making it like physics in being mathematical. For, obviously, there would be two things wrong with such a reading. First, Fisher is often concerned as much with contrasts between evolutionary biology and mathematical physics as he is with comparisons. Second, it is not merely the mathematical form of equations, measurements and derivations in mathematical physics that Fisher sees as enlightening for evolutionary theorising. It is, more often and more fundamentally, the conceptual and, indeed, philosophical implications of the content of the physics itself, rather than its mathematical form.

It is in fact interesting to note that Fisher's mathematics, that is, principally, his mathematical statistics — including what he saw as its lessons for the understanding of statistical inference and, beyond that, inductive logic — has little discernible relevance to his thinking about genetics and evolution. This may be surprising. It would be reasonable to conjecture, for instance, that his early work on the statistics of small samples might somehow have conditioned his approach to the theory of natural selection. However, it is hard to find any evidence for this. In fact, when Fisher wrote on the work of his hero Gosset ('Student'), he discussed Gosset's pioneering insights into small sample statistics and later noted that Gosset was also right-minded, by Fisher's lights, on evolution, being a good selectionist; but Fisher made no link between these two elements in Gosset's thought.[40] It would be rewarding to find links, in Fisher, between his mathematics and his biology; links, that is, other than those — such as the obvious one provided by statistics itself — that are mediated by topics in physics. However, so far they have remained elusive, and have therefore to be left aside here, as belonging among future possibilities rather than among biographical results already achieved.

It will be apparent already that any account of Fisher's eventual philosophical synthesis of physics, biology and eugenics could equally well take as its guiding theme his Christianity, his Darwinism, his theory of civilisation or his statistical view of nature and man. For our purposes here, however, one choice recommends itself as especially appropriate. That is Fisher's indeterminism. In particular, this choice is appropriate because Fisher himself was most explicit, in his philosophical writings, in presenting his indeterminism as informing his outlook on everything else. We do well, therefore, to examine how and why he could see it functioning in such a comprehensive way in his philosophical life.

Before embarking on that examination, it will be as well to have some sense of Fisher's style and sympathies as a philosophical thinker. One distinguishing feature emerges, for example, from any glance at his intellectual biography. He combined within him both what we may call scientism and what we have to call romanticism. By scientism here is meant a commitment to give science an ultimate authority on all subjects: in a word, to set no limits to the domain nor to the authority of science. On the other hand Fisher has been called a romantic, and rightly, because of this taste, both in thought and deed, most obviously in his twenties, for pursuing flights of fancy and fantasy so infused with sentimentality, nostalgia and idiosyncrasy that anyone thinking of himself or herself as a realist and a rationalist child of the Enlightenment would have to view Fisher's life and thought as distinctly alien.

Two youthful compositions can take us from an exemplar of Fisher's scientism to a no less exemplary instance of his romanticism. Already showing that drive for generality in any scientific principles he embraces, Fisher opened his 1912 paper on "Evolution and Society" by declaring that Darwin's theory requires very little to be assumed about the "nature of species". The "same process" of selection goes on in "languages, religions, habits, and customs, rocks, beliefs, chemical elements, nations, everything else" that may be "stable" or "unstable". For natural selection all that is needed is that the "suitable to survive" shall do so, while the "unsuitable, unstable" do not. So, the selectional history of the habit of smoking or bridge playing or family prayers — Fisher's own examples — can be dealt with in the "same very general manner" as the evolution of the worms parasitic in some parrots. Fisher does not go on in fact to deliver on this promise as such, but devotes the rest of the paper to intersocietal selection in man; concluding that this selection tends to favor social specialisation and regimentation, but that we may "still hope that magnificent qualities and capabilities of the best type of man will render specialisation unnecessary" and that "the small, spirited nations were right" to believe "liberty was better than regimentation".[41]

There may be no sure echo here of Nietzschean romanticism. But Fisher was already a devoted member of a circle of friends who gave each other names taken from Nietzsche's *Thus Spake Zarathustra*; and

within three years he had published a piece on 'The Evolution of Sexual
Preference', in the *Eugenics Review*, that climaxed in a final three
sentences (crossed out later by Fisher in his copy) where "we pass, like
Nietzsche, beyond Good and Evil", with morality ceasing to be "arbitrary and dogmatic" but taking "its place as a particular formulation of
the requirements of the Highest Man — of our ultimate judgments of
human value". This finale is reached by starting with a defence of
Darwin's theory of sexual selection, and then arguing that all aesthetic
and ethical generalisations have natural evolutionary origins in sexual
selection, so that aesthetics and ethics themselves are grounded in the
judgments made by individuals in their choice of mates. Eventually,
there arises a struggle between ethical and aesthetic valuations, with
ordinary people rating beauty superficial and moral worth fundamental.
But "in the deepest minds" the idea of beauty links itself to "nothing
less than the mystical appreciation of human personality", the "highest
plane", and "the source" for all other, lower valuations. It is, here, then,
that Good and Evil are surpassed.[42]

So, whatever else Fisher's philosophy has to do for him, it has to give
him a universe where liberty and beauty can hold a place in the life of
Christian and a eugenist. By 1932, certainly, and very likely several
years before that, Fisher had seen his way to securing such a philosophy, by starting from what he took to be a finding of science itself:
namely, that the Boltzmannian way down and the Darwinian way up
can, indeed should, be taken as freeing us from determinism and all its
implications.

9. THE BIOLOGY AND PHILOSOPHY OF AN INDETERMINIST

It is essential, then, to appreciate how scientistic is Fisher's very
approach to philosophy. His 1934 paper on 'Indeterminism and
Natural Selection' is quite explicit: many older philosophical positions
have been discredited by the progress of science. The Greeks constructed plentiful philosophical theories, but without sufficient knowledge of empirical science these philosophical exercises of the imagina-
tion were doomed to later implausibility. In modern times philosophy
has taken empirical science into account; but there is often a lag, in that
doctrines live on in philosophy that are no longer sanctioned by the
best empirical sciences. So it is with determinism, Fisher argues. He
goes to some length to argue that seventeenth-century science may have

made determinism reasonable, but that developments in nineteenth-century science have made it no longer inevitable or even credible. Once again it is the Boltzmannian and Darwinian developments that are jointly decisive. "The historical origin and experimental basis" of "physical determinism" show that "this basis was removed" with the "kinetic theory of matter", while its "difficulties" grew with the evolutionists' admission that "human nature, in its entirety, is a product of natural causation".[43]

Fisher is explicit that no twentieth-century views on "indeterminacy" as discussed in quantum physics, add any new extra argument, only a welcome supplement to the nineteenth-century foundations for indeterminism. The reason is that the indeterminism Fisher defends holds only that reliability and predictability in aggregates, such as gases, require no determinism — only numerousness and independence — in the motions of the component elements. Nor is there any evidence, from other grounds, of component determinacy in nineteenth-century science. What twentieth-century physics has done then is merely, once again, to find no such evidence since. Appealing to generality, as ever, Fisher can, therefore, argue that indeterminism is the more general assumption, in that it avoids the additional special presuppositions about component elements that determinism makes.[44]

So enamored is Fisher of his indeterminism that he is prepared to present it as serving a vast range of philosophical needs, without entailing any serious drawbacks at all. Thus, in the eighteenth-century man, as freely and rationally choosing and acting, seemed beyond all possible scientific explanation because scientific explanation was thought to require deterministic causation. Conversely, then, when evolutionary biology brought man within science the plausibility of the old deterministic exclusion of man was challenged. With an indeterministic view that does not oppose the laws of nature to the laws of chance, natural and human science, including social science, can be unified. Marriages remain free, individual human choices and actions, but the statistical lawfulness of variations in aggregate marriage rates shows that in this, as in everything else, man and society come within a single scientific enterprise seamlessly joined with natural science.

Fisher can accordingly reduce to one issue the implications for social policy of the great shift from Laplacean deterministic science to Boltzmannian-Darwinian indeterminism. The possibility of science requires only numerousness and independence in the component elements in the aggregate being studied. Human societies are all obviously numerous enough in the components; but the issue of independence remains to be understood. Social organisation and public opinion are the two main features of society limiting independence of individual actions. However, Fisher's eugenics is not frustrated by these features. Indeed it can be made effective by being accommodated to them. For a start, heredity — the transmission from parents to children of the relevant traits — proceeds apart from any limits on independent action set by social organisation or public opinion. Now, in addition to heredity, the eugenist needs only to consider one other item: social promotion or the assignment of individuals to occupational grades and any correlation it may have with heritable fertility or infertility. The social promotion of infertility, this correlation, appears from historical sociological research to be an inevitable feature of every civilised society where social class is present, wealth is a factor in social promotion and children are an economic burden. Any measure that countered this last feature would, then, be effective in reversing the aggregate correlation by providing incentives for numerous social promotions to be won by more rather than less fertile individuals.[45]

The great nineteenth-century developments in kinetic theory show, therefore, that indeterminism is entirely consistent with the orderliness in the natural and social worlds and with success in the quest for knowledge, including causal knowledge, of those worlds. As for causation, far from diminishing this concept, indeterminism enhances it. With determinism everything is determined all the way forward from some indefinitely remote moment in the past. So, there is no room to think of subsequent events as causing anything that was not already determined. By contrast, with indeterminism, events can be understood as originating new lines of causation, and of making a difference by making things happen that would not otherwise have done so. The asymmetries in human and animal remembering, striving and aiming make no sense with determinism. If the future is as fixed as the past, memory should not refer uniquely to the latter; nor striving and aiming to the former.[46]

By enriching the concept of causation as it does, indeterminism clarifies the concept of creativity and so the relations between science, on the one hand, and art, ethics and religion on the other. Within science causation is creative in a strictly scientific sense. In a game of chance, we can predict all the possible forms of the result, stating in advance their probabilities of occurrence, without foreseeing just which will occur. Likewise for any event, it can be creative if it can be causal, in that it can be the one that makes the difference in bringing about a new event what could not be predicted from earlier events. Beyond causality, however, the concept of creativity may imply an emotional stance. Being merely new in time, like a new penny, is not enough to exemplify creativity in this fuller sense; there must also be novelty in the "nature and potentialities" of what is produced. "This is intended when we apply the word to the work of a scientist, or an artist; that his work matters in itself, and to the future of his art or science". To be creative in this sense work "must have value, intellectual or aesthetic, moral or social value; consequences which excite wonder, or admiration".[47]

Fisher's view is, therefore, that any truly causal process is creative in the scientific sense; it may but need not be creative in the full sense. One cannot complain that the causal process invoked by some scientific theory is not creative in the full sense. That depends on our feelings about the consequences.

10. THE DISPELLING OF MECHANISM, MATERIALISM AND PESSIMISM

On this ground, Fisher welcomes the insistence of the philosophers Henry Bergson and Jan Smuts on talking of "creative evolution"; and of Alfred North Whitehead's (one of his teachers in mathematics) emphasis on nature as a creative process. However, where Fisher disagrees with Bergson and Smuts is over their view that Darwinian evolution by natural selection is not creative in the appropriate senses. Bergson has denounced Laplacean mechanism as inconsistent with our very consciousness of temporal duration. And he has argued that Darwinian or Lamarckian evolutionary biology is no less mechanistic, and so no less inconsistent with what is most indisputable in our experience. But, Fisher holds, Bergson has mistaken what it is that we need to be liberated from in Laplacean science. It is not mechanism, whatever that might be, but determinism. Make the shift to Boltzmann and Darwin properly understood, and one has all the philosophical dividends that Bergson is rightly seeking, but seeking in the wrong place. So Bergson's version of evolution, which invokes an *élan vital*, a vital impetus peculiar to living things, is philosophically unnecessary as well as unacceptably mysterious and empirically unsound in its claims about the role of mutations in evolution.[48]

Here, as always, Fisher rejects a choice between mechanism and vitalism. The distinction between deterministic and indeterministic causation can be defined with rigor, Fisher insists, but no one has satisfactorily distinguished a "mechanistic from a vitalistic organism". The only issues Bergson can ultimately raise are, therefore, the ones that are all resolved by an indeterministic view of science including Darwinian evolutionary theory, with its new, genetical theory of natural selection.[49]

Likewise with Smuts's call for holism. Smuts demands a holistic, creative process in the animal or plant germ, somehow eliciting the mutations needed for evolution. But Fisher, having praised the "wisdom and width of his more essential views" and the "religious feeling" he brings to his interpretation of evolution, urges that Smuts has no need to find holistic, creative processes inside zygotes. Properly understood, says Fisher, natural selection puts the creative causation at the outside boundary of every organism. For it is in the interactions of organisms with their environments, and so in the successes and failures of all the functions of all their organs, that creative causation is found. For conscious beings, this boundary is where consciousness itself resides, on the boundary between our inner subjective life and the outward objective sources of sense experience. More generally, the theory of selection seems "holistic" to Fisher, in its emphasis on the "mutual reaction of each organism with the whole ecological situation in which it lives — the creative action of one species on another".[50]

Faced with the charge that Darwinian evolutionary biology is materialistic, pessimistic and somehow "soulless" in its vision of nature, Fisher replies that he can not agree; because he sees nature as full of creative activity and of signs that attempts to lessen evil and promote progress can succeed. While Lamarckism reminds him of the doctrine of salvation by faith, in that strivings are decisive, Darwinism joins works to faith in making actual achievements, doings and dyings, decisive. It is then "hard" to see "anything unedifying or disquieting" in a theory of evolution that rests on the actual "performances" of organisms. As for evil, it is, on Fisher's view, "relative" in that "it changes its nature with evolutionary progress and with the changing structure of human society".[51] The torture of prisoners and the wreaking of man-made ecological disasters rightly concern us more now than the sins proscribed in the Old Testament. As Fisher might have argued, today, the philosophy of an indeterminist Boltzmannian and Darwinian scien-

tist coheres well with the ideals of Amnesty International and Green-
peace.

But even if indeterminism can discredit all charges of mechanism,
materialism and pessimism, is there not still a defense needed of our
own free wills as individuals? Sometimes Fisher contents himself with
merely asserting that we know our wills to be free in the only senses
required for a defense of active policies for progressive social change.
However, he had struggled with free will in letters to Leonard Darwin
and returned to it in correspondence with the neurophysiologist C. S.
Sherrington later. In 1934, he had argued that a mere illusion of free
will could have no survival value and so was inconsistent with crediting
all our human nature to natural selection. But that thesis had still left a
difficulty: namely, "to allow the evolutionary process, which depends
upon the permanent and therefore deterministic properties of genes, to
take any part in the development of such a capricious quality as the
possession of powers of individual choice". Here is Fisher's "attempt to
set out a possible relationship" between the evolutionary process and
individual choice:

> The development of a given genotype (even in given environmental conditions) is
> indeterminate in that undirected chance happenings intervene at all stages, each such
> event having perhaps permanent or increasing consequences, as development proceeds,
> on the integration of the nervous system and the formation of character.
>
> Individual action, e.g., choice, is always in part predetermined by the genotype, in
> part by the subsequent effects of physically fortuitous developmental happenings in the
> past, and in part undetermined and ascribable to fortuitous contemporary happenings.
>
> Both the course of development, and the instantaneous state of the nervous system,
> are such as to amplify the effects of initially minute (quantum) events, so as to have
> molar consequences.
>
> This general principle of amplification has been of importance to survival, in some
> way at present obscure, perhaps connected with the organization of the whole bodily
> mass into individual unity, perhaps in orienting its reactions towards the future (as
> purpose or intention), and has evolved to its present high degree by reason of its
> survival value. It, though not the particular modifications which it favors, is determined
> by the genotype.
>
> It is open to a man, religiously inclined, to assert that the primary elements of
> indeterminacy in development and choice are fortuitous only in the physical sense,
> being in reality divinely guided, much as the apparatus of games of chance were
> regarded as guided by the Goddess Fortuna.[52]

This is Fisher in 1947. In 1922 he had been one step ahead of
Wright in population genetics; but in the philosophy of the freely
choosing mind of the individual, Fisher was unwittingly retracing a path

that Wright had already taken before they had ever met. This makes an appropriate moment, therefore, to switch to the young Wright.

11. WRIGHT: ADJUDICATIONS FOR SCIENCE AND WITHIN SCIENCE

It would be tempting, but also misleading, to introduce the young Wright by looking at those biographical landmarks that match up best with corresponding moments in Fisher's early life. It would be misleading because their early lives were too different in decisive ways. To be sure, Wright, like Fisher, read Darwin at school and apparently became from then on in favor of Darwin's theory of gradual evolution by natural selection. But this parallel with Fisher is a limited and superficial one, in that Wright was eventually to owe his views on the place of selection theory in evolutionary biology to educational and research contexts that had no equivalent in Fisher's life, a life it has to be remembered that included no formal instruction in biological science after high school.

Wright was many years in the making as a qualified zoologist. The years (1906–1911) at Lombard College, ending with a bachelor's degree, took in a wide sampling of biological subjects and various texts on evolution. A year (1911–12) at the University of Illinois was preceded and followed by summer visits to Cold Spring Harbor. So, when Wright went to Harvard's Bussey Institution in 1912, to work with Castle toward a PhD in genetics (awarded in 1915), he was extensively prepared in the relevant disciplines, but had had no occasion to work out explicitly and precisely where he stood on the issues dividing evolutionary theorists at the time.

However, it seems that there was never for Wright a choice to be made between Mendelism and Darwinism, or between Mendelism and biometry. He grew up, so to speak, seeing the three as consilient; this consilience being more easily assumed in the U.S., where the English opposition between the Mendelian Bateson and biometrician Pearson had no close equivalent. At the Bussey Institution, such a consilience was especially strongly represented in the views of Castle and East.

The question arises, then, as to whether eugenics formed for Wright, as it did for Fisher, a fourth commitment to go with a youthful embracing of Mendelism, Darwinism and biometry. Wright's attitude to eugenics at this time is not easily discerned from direct documentation.

However, it seems most likely, from the stance he would later adopt, that it was never an enthusiasm, nor something, conversely, that he was bent on opposing. More generally, Wright does not seem to have actively aligned himself with any political doctrines or groups. The overall impression one has is that he felt most comfortable with what would then have been, for an academic son of an academic family, broadly the middle ground. For a time, when Wright was an undergraduate, his father evidently moved out of the middle ground toward socialism, as upheld by Eugene Debs; but the son was less enthusiastic.[54]

Likewise with religion; it seems, although here we have even less direct documentation, that there were in Wright no strong feelings, religious or hostile to religion, for him to bring to his science in general, or to his understanding of evolution in particular. And yet, before he had finished his first year at Harvard, Wright was reading Bergson's *Creative Evolution* — presumably in the English translation published the year before, in 1911 — and being prompted to embark on a philosophical quest that eventually led him to construct a comprehensive metaphysics of mind and matter of his own, a dual-aspect panpsychism, wherein the ultimate, inner reality of everything from ions to men, is held to be mental, while science studies only those external appearances and observable actions that constitute the material aspects of things.[55]

What can be said, therefore, about the motivation of this philosophical quest, if it has no obvious political or religious rationale? Can we anticipate, here, what a closer analysis of Wright's intellectual biography might reveal, so as to identify those themes that must run through any discussion of the relations between his science and his philosophy? Such a task is harder in Wright's case than Fisher's. However, there is one theme that it may be useful to announce in advance: the theme of adjudication. For Wright's philosophical quest started from a quite general issue: the conflict, especially as Bergson had identified it, between the Laplacean determinism presupposed by much scientific thought, and the phenomenon of consciousness itself. Now, Wright saw this conflict as a quite general one. Although biologists were peculiarly involved with the relation between mind and matter — unlike physicists who could usually ignore minds and psychologists who could sometimes ignore matter, as Wright reflected — the only resolution of the conflict that was satisfactory for a biologist would be one that comprehended all natural science. What is more, Wright's eventual resolution of it had as an explicit corollary that natural science takes no account of the inner, ultimate mental reality of things, and can, therefore, proceed deterministically — at least in the sense of determinable probabilities — without conflicting at all with any convictions that may be held about consciousness and minds. Unlike Fisher's indeterminism, therefore, Wright's panpsychism is not designed to show that the content of the best science is such as to satisfy any legitimate metaphysical demands that we may make of it. Quite the contrary, for Wright's panpsychism is designed to show that science, limited as it is to the material aspects, can not be expected to meet any metaphysical demands. Only the philosophy of panpsychism itself can meet those demands.

There is, then, in Wright no equivalent to Fisher's vindicative attitude toward Darwinism, the commitment, that is, to vindicate Darwin's theory both scientifically and philosophically. Within evolutionary biology, Wright sees himself as an adjudicator. Where Fisher seeks to rethink — and vindicate — Darwin in the light of the new genetics, for Wright it is not Darwin, or even natural selection, but rather evolution itself that has to be rethought in the light of the new genetics. That rethinking is what the statistical theory of population genetics has to contribute. In such a rethinking, the claims of rival theories — De Vries's mutationism, Wagner's isolation theory, Eimer's orthogenesis and Darwin's selectionism — have to be adjudicated, by establishing under what conditions this or that trend, agency or tendency predominates in conditioning the course of evolution. For this reason, Wright did not write under any such title as Fisher gave his book, but addressed himself to the more comprehensive topic: 'Evolution in Mendelian Populations' and 'The Roles of Mutation, Inbreeding, Crossbreeding and Selection in Evolution'.

The way Wright carried out his adjudicational business regarding evolution was conditioned, above all, by a feature of his career that, once again, has no equivalent in Fisher's. Fisher had gone straight to his vindicative business, in the sense that he had always brought his developing knowledge of the new genetics — knowledge of genes and mutations and so on — directly to bear on the theory of evolution by natural selection. Wright, on the other had, was to come at evolution through the intermediary subject of livestock breeding, its theory and practice. Before rethinking evolution in the light of the new genetics of the first twenty years of the century, he had worked at rethinking livestock breeding in that light; only subsequently did he elaborate his theory of evolution as a generalisation and accommodation of his livestock breeding conclusions to the resolution of what he saw as the main challenges in understanding evolution itself. Many years later, Wright would indicate that he saw cultural evolution in man as analogous to organic evolution, so that his shifting balance theory of organic evolution was to have its analog in understanding cultural evolution. With Fisher, it was enlightening to ask what he saw as common to all processes of selection. With Wright, one needs to ask what it is that is seen as common to livestock breeding, organic evolution and cultural evolution. The answer is explicit and reiterated by Wright himself: the decisive feature is cumulative change, that is change involving persistence as well as variability. We have, therefore, to think of Wright as always theorising by adjudicating among diverse proposals concerned with making intelligible — to use his phrase — processes of cumulative change. Such phrases remain mere phrases, needless to add, until associated with particular domains. We should look without delay, then, at Wright's livestock breeding work.

12. FROM BREEDING TO EVOLUTION

There is now a standard way to reconstruct the origins of Wright's shifting balance theory of evolution, and although there is no reason to question it one does need to ask precisely what it can and can not do for us biographically. The standard reconstruction traces directly to Wright's own retrospective insights. These insights, unlike many scientists' autobiographical reflections, are unlikely to be seriously mistaken for two reasons. First of all, Wright's habit, throughout his life, was to refer back constantly to his earliest publications and his earliest researches. He was always keeping track, far more than most scientists do, of where he had been and how he had reached whatever was his present position. Second, where Wright's memories have been checked against independent evidence they have shown themselves remarkably accurate. We can therefore have reasonable confidence concerning his account of what was decisive in forming his views on evolution.

On one point, however, we have to stay critically alert. Wright's own most detailed version of that account was given originally as a talk, and as a talk before a meeting of the American Society of Animal Science, formerly the American Society of Animal Production. It was then addressed to an audience with a special interest, in every sense, in livestock breeding. And sure enough, Wright called the talk: 'The Relation of Livestock Breeding to Theories of Evolution', although it is his own theory alone that is discussed in any detail. Now, in that talk Wright distinguished four lines of his early work, prior to 1925, that were decisive for his theory. Given the title of his talk, one might think that Wright was recalling all four as somehow coming out of his livestock breeding studies during his years (1915—1925) with the U.S. Department of Agriculture. But, as Wright himself makes plain, two of the four trace to his years at the Bussey and so before he had any special, professional concern with breeding. A reconstruction of the whole story would have, therefore, to ask how the Bussey years' work conditioned the U.S.D.A. years' thinking and, then, how the work of both periods was drawn on in the typescript exposition of the evolution theory composed in 1925, but not published until 1931 in a modified form.[56]

Consider the two Bussey conclusions, as set out in Crow's lucid retelling of Wright's recollections. First, Wright was convinced by Castle's selection experiments that selection was effective in producing a divergence between hooded rat strains that were almost completely white or almost completely dark; but he was also impressed by the need to stop the experiment eventually on account of the infertility produced. The conclusion seemed to be that mass selection can be effective in changing one trait, but that intense selection also brings other, unwanted traits. Second, Wright was convinced, by his own doctoral study of hair color and pattern in guinea pigs, that genes in combination often had unpredictable effects; and that selecting for individual traits may not produce the desired combination. Obviously, then, Wright entered the U.S.D.A. and took up his breeding researches already convinced that mass selection of individual traits was often efficacious but also not reliably optimal.[57]

Consider next, then, the two U.S.D.A conclusions, again as set out in Crow's version. First, the study of inbreeding and crossbreeding in guinea pigs confirmed that inbreeding leads to a decline, but it also increased random differentiation among different lines and fixed distinct combinations — not only of superficial traits such as colors, but of internal physiological traits. Among those distinct combinations were occasional favorable ones. Second, the history of livestock breeding, of

Shorthorn cattle especially, showed the advantage of combining selection with inbreeding within herds followed by the exporting of sires from superior herds to others, with a consequent benefit to the whole breed. Now, one can accept that — as Crow puts it — by replacing "guinea pigs" and "cattle" in these four conclusions with "natural populations," one has the essential core of the shifting balance theory.[58] However, what still remains to be asked is why Wright saw the problems posed for breeders as sufficiently like the problems addressed in evolutionary theory, that a solution could be transposed in this way. It is here, as already indicated, that we have to concentrate on Wright's preoccupation with the issues he introduces through such telling terms as *cumulativity, plasticity, balance, levels* and the *interpolation* of a process of mostly random differentiation of local strains between the primary gene mutations and the control of major evolutionary trends by natural selection.

To look ahead, these various issues come down to two main clusters that can, perhaps, be seen as ultimately one. A first cluster is introduced right at the opening of the third and final section of the 1931 paper, the section — 'The Evolution of Mendelian Systems' — that attempts to draw conclusions about evolution from the findings of the previous two sections — those on the 'Variation of Gene Frequency ' and 'The Distribution of Gene Frequencies and Its immediate Consequences'. In attempting to draw such conclusions one may assume, says Wright, that causes making for variation are favorable, those reducing variation unfavorable. However, evolution being cumulative change, rather than mere change, "fixation in some respects is as important as variation in others". Livestock breeders, says Wright, liken their work to modelling in clay, speaking of moulding the type toward the chosen ideal. The analogy is a good one, suggesting as it does that in both cases "a certain intermediate degree of plasticity" is required.[59]

A second cluster was often insisted upon when Wright contrasted his views with Fisher. Wright's early studies of factor interaction concluded that genotypes relate to phenotypes "by a very complex network of biochemical and developmental reactions", with each character usually affected by many gene substitutions, each substitution having many pleiotropic effects and the intervening processes involving nonadditive interactions. This "viewpoint" thus contrasts with the "common treatment of organisms as mosaics of unit characters (Fisher's norm)" when considering their evolution. So, on Wright's viewpoint, "evolution becomes a much more intelligible process if based on natural selection among interaction systems rather than among alleles at each locus separately". Without strong linkage, the allelic selection is all that is possible with natural selection among individuals in a panmictic population where Fisher's fundamental theorem would apply, because, here, recombination quickly lead to the dissolution of any particular gene combinations. The more intelligible process is, however, possible with inbreeding and with selection at two levels. With the Shorthorn cattle there had been selection by particular breeders in building up their herds, and by breeders in general among herds as sources of sires. The first favors "the allele at each pertinent locus that gives the most favorable effect on the average of all combinations with such alleles at other loci". This, says Wright, is *"genic* selection"; and the resulting patterns are "fixed, more or less, by close inbreeding". Selection among such herds as sources of sires is "selection among the diverse interaction systems that happened to have been arrived at". This, he says "is *organismic* or (genotypic) selection". It was recognising "that the two-level process was much more efficient than mere individual selection" that led him to see "whether an analogous two-level process might not occur in nature".[60] Wright does not make his point explicitly in this way, but we can say, in summary, that the interpolation, in the conception of the organism, of complex gene interactions, mediating between gene substitutions and phenotype character differences, required in the conception of population structure, the interpolation of inbreeding, random drift and selection in local races, subgroups of small population size, between the reproduction of individuals and the changes taking place in the species as a whole.

13. BALANCE, LEVELS AND THE INTELLIGIBILITY OF EVOLUTION AS CUMULATIVE CHANGE

As an exegetical policy it proves most instructive to concentrate on the 1931 paper. It is not that Wright's shorter, less demanding and more familiar 1932 paper is misleading in its representation of his views.[61] It is, rather, that the 1931 paper is more explicitly related to those general issues that provide the common grounding of his thinking about livestock breeding and evolution. The analogies and metaphors elaborated through the 1932 device of the adaptive topography allowed the exposition of the shifting balance theory to be sundered from those issues, and expressed briefly in pictorial epitome without reference to the framework assumptions that are so prominent in 1931.

Of the more than five dozen pages of the 1931 piece, only a dozen and a half — all but two of them at the end — are explicitly devoted to evolution. It is easy enough, then, as well as indispensable for our purposes, to see what are Wright's principal proposals. A fundamental lesson that he wants to instil is this: "The problem is to determine how an adaptive evolutionary process may be derived from such unfavorable raw material as the infrequent, fortuitous and usually injurious gene mutations". This is the problem because the "basic cumulative factor in evolution is the extraordinary persistence of gene specificity", while the "basic change factor is gene mutation, the occasional failure of precise duplication"; and because the conclusion from the "present status of genetics" is that "any theory of evolution" must be based on the properties of Mendelian genes together with the "statistical situation in the species". That these two resources are decisive follows from "the fact that the evolutionary process is concerned, not with individuals, but with the species, an intricate network of living matter, physically continuous in spacetime," and with its responses to external conditions, which relate to "the genetics of individuals only as statistical consequences of the latter".[62] For this restriction to statistical consequences follows, Wright argues, from the discrediting of Lamarckian and other theories that would relate changes in the environment to genetic change in the species through their direct physiological consequences.

A constrast with Fisher is already apparent, then. Where Fisher stresses that heritable variation has turned out to be, in the new genetics, exactly what a selection theorist needs to free him from all the difficulties he had with blending inheritance, Wright takes the wider view: as an evolutionary theorist, a theorist of cumulative change, including adaptation, Mendelian mutations are, in and of themselves, distinctly unpromising. The study of the statistical situation in the species, to use his phrase, has, therefore, a big challenge to overcome.

Wright proceeds to clarify and meet this challenge with argumentation that has no equivalent in Fisher. More particularly he begins his argumentation by giving himself resources that have no equivalent. Their significance is well worth emphasising, therefore, since they can provide light on Wright's entire caste of mind as an evolutionary theorist.

The resources relate, once again, to the theme that cumulative change requires factors that make for persistence as well as factors making for innovation. No understanding of Wright's shifting balance theory can ignore this theme, precisely because the balance proclaimed in the very name of the theory is a balance between persistence and innovation.

Wright lists nine pairs of such factors in two columns: 'Factors of Genetic Homogeneity' and 'Factors of Genetic Heterogeneity'. Some pairings are straightforward enough, such as the first — gene duplication and gene mutation — and the fifth — linkage and crossing over. With others there are complications. Environmental pressure (s, for selection, in the equations) is a homogenetic factor opposed to individual adaptability, because individual adaptability allows similar genotypes to develop into dissimilar phenotypes and so to reduce the diminution of variance due to environmental pressures exerted as selection. But then individual adaptability also shows up as a homogenetic factor opposed to local environments of subgroups (s_1 in the equations); for it also allows for dissimilar genotypes to develop into similar phenotypes even in differing local environments. Clearly, therefore, for some of these factors, whether they are homogenetic or heterogenetic will depend on the circumstances and on what else is going on.[63]

Obviously, too, we have here a conceptual framework that differs totally from Fisher's with its division of all causation into two categories: the reliably, if not strictly invariably, counterentropic (natural selection alone) as opposed to all the rest, which are entropic. This contrast with Fisher is confirmed when we turn to the second of Wright's resources. For Wright goes on to explain the role of these factors in evolution by considering, in certain cases, the difference that was made to all subsequent evolutionary processes by their original introduction in the history of life. Thus he considers the evolutionary process in, first, viruses where the gene is the organism and, second, bacteria and blue green algae, which have, as he holds, gene aggregates but no mitosis or meiosis, and where, therefore, "the conditions are not favorable for an extensive cumulative process". Not surprisingly, Wright, citing East, urges that the "most important factor in transcending the evolutionary difficulties inherent in the characteristics of gene mutation" is "biparental reproduction". But while this provides a rich field of variation, "by itself it provides rather too much plasticity"; for along with the adaptability it gives to the species comes the consequence that "a successful combination of characteristics is attained only to be broken up in the next generation" by meiosis itself.[64]

Now, "the principle that a balance between factors of homogeneity and of heterogeneity" is more favorable for evolution than either by itself is well illustrated by the effects of an alternation of a run of asexual generations with an occasional sexual generation. This works very well both in nature, in many species, and as a technique deployed by plant-breeders, because beneficial combinations generated sexually can be preserved and multiplied asexually. However, asexual reproduction is often absent, especially in higher animals. The consequence of its absence is definitive of the very "purpose" of his paper, Wright explains, which is "to investigate the statistical situation in a population under exclusive sexual reproduction" so as to clarify "the conditions for a degree of plasticity in a species" that makes "the evolutionary process an intelligible one".[65]

This characterisation of the very purpose of Wright's paper is accordingly followed by an examination of the various homogenetic and heterogenetic factors as they influence any population. Examining their consequences in large populations, small populations and medium ones, Wright argues that intermediate population size is optimal, large populations being subject mainly to slow and often to reversible changes with selection, while small ones are too liable to decline and extinction from random fixation of disadvantageous genes. By contrast, with a population of intermediate size, a "continuous and essentially irreversible" change seems ensured "even under completely uniform conditions", the direction being "largely random over short periods but adaptive in the long run".[66]

This conclusion sets the stage, in turn, for the analysis of what happens in a large population divided into partly isolated subgroups of small size. Here, even though the subgroups are small the results are much as they are for populations of intermediate size. There is a "partly nonadaptive, partly adaptive radiation among the subgroups". Those coming up with "the most successful types" would "presumably flourish and tend to overflow their boundaries while others decline, leading to changes in the mean gene frequency of the species as a whole". And, in this case, "the rate of evolution should be much greater" than is the case of a single undivided population. This last comment shows, then, that Wright's notion of what is optimal includes not only the constraints of cumulativity and adaptation but also of quickness in achieving perma-

nent, adapted productions. Once again, it is plain the these constraints
on any theory of evolution are carried over by Wright from his livestock breeding analyses.[67]

It is plain, too, that the smallness of the partially isolated subpopulations is decisive for Wright for reasons that have to do both with
random drift, in the sense of fortuitous or random or indiscriminate
differential reproduction, and with inbreeding, the breeding together of
relatives. The mathematical treatment of random drift and inbreeding
tends to assimilate the two together, but their significance for Wright's
overall theory can be obscured by that assimilation. Inbreeding is a
homogenetic, fixational factor in any one population, allowing for
complex interactive gene systems to be held together. Random drift, on
the other hand, as is emphasised in the 1932 paper, counteracts the
tendency of the intrapopulationally homogenetic factor, selection, to
hold the population at the nearest adaptive peak; for drift allows the
population to cross an adaptive saddle and so come within the selectional influence of a higher peak.

These complex reasons for Wright's emphasis on the advantages of
smallness in subgroups are worth noting here, because in the ensuing
debates between Wright and Fisher and, more broadly, among their
respective followers and champions, there was a tendency to concentrate on only one element in Wright's overall proposal: namely, that
drift will be tending to cause nonadaptive divergence among subgroups
and so, too, perhaps, eventually among congeneric species resulting
from any eventual speciations. A fuller understanding of what Wright
was about requires concentration, as always, on his preoccupation with
persistence and variation as conditions for cumulative change. The
precise wording of familiar sentences from the closing summary
paragraph of the 1931 paper can put us on to a better reading of the
paper as a whole. As "a process of cumulative change" evolution
"depends on a proper balance of the conditions, which, at each level of
organisation — gene, chromosome, cell, individual, local race — make
for genetic homogeneity or genetic heterogeneity of the species". In a
large, subdivided population, there is "a continually shifting differentiation" among the local races "intensified by local differences in selection
but occurring under uniform and static conditions" and inevitably
producing "indefinitely continuing, irreversible, adaptive and much
more rapid evolution of the species".[68]

14. THE BIOLOGY AND PHILOSOPHY OF A MONISTIC DUAL-ASPECT PANPSYCHIST

Wright's 1931 paper included a short subsection, of a single paragraph, on " 'creative' and 'emergent' evolution", where he explained how he saw his scientific account of evolution relating to his philosophical views. The discussion is so abbreviated, however, that to go to it directly is to risk missing the items that were decisive for Wright's intellectual development. Fortunately, in three other papers, written in 1953, 1964 and 1975 respectively, he dwelled autobiographically on the origins of the position encapsulated in 1931: 'Gene and Organism', 'Biology and Philosophy of Science' and 'Panpsychism and Science'.[69] What these three papers make plain is that there are two ways to go astray in understanding the relations between philosophy and biology in Wright's life and work. The first would be to read his views within biology as somehow deriving as a consequence from more comprehensive philosophical views. This would be a mistake because, as Wright himself insists, his philosophy was designed to leave his biology following ideals of method, evidence and explanation that were already commonly if not universally accepted among working scientists. However, the second mistake is to infer from this that Wright's philosophical concerns, the questions he asked, the answers he considered, the reading and reflecting he did, were without influence in conditioning his life as a biological theorist. For in decisive ways it is pretty certain that they were influential. In making such a claim for this kind of influence, it will be as well to admit to yet another underdetermination thesis. Just as scientific theories are held by many to be underdetermined by any factual evidence for them, so it seems are scientific and philosophical views by each other. Whatever influences run back and forth still leave those views less than strictly determined. But to acknowledge that limitation is still to find room for a biographically defensible case for conditioning influences.

Consider next, then, Wright's recollections of what three books did for him in the years from 1912 to 1914. Before this he had accepted that science requires a rigid determinism of the Laplacean kind, but he had worried about where consciousness could find its place. This determinism was disturbed by reading Bergson in 1912, "but not for long", in that Wright was to return to determinism "in practice but with a radical revision of the philosophical implications of science".[70] How, then, was this return made along with this radical revision? A decisive step came on reading, in that same year, the biochemist Benjamin Moore's book, *The Origin and Nature of Life* of 1912, with its suggestion that cells and higher organisms and societies could be seen as extensions to the series running from atom to molecule and to colloid. This Wright accepted, but still a "dilemma" remained: "absolutely deterministic laws of physics at one end of the scale, consciousness and apparent freedom" at the other end.[71] The resolution of this dilemma followed a reading in 1914 of Karl Pearson's *Grammar of Science* (the second edition of 1899). This resolution involved two moves, one that was provided directly by Pearson, and another that was definitely not, as Wright himself emphasises. The first was the move to considering laws of nature not absolutely, as "part of the eternal structure of the world", as Wright had hitherto considered them, but "as merely condensed statistical descriptions of how things are observed to behave". They would be, then, no different in kind from "statistical laws of voluntary human behavior such as the law of supply and demand". Wright, moreover, accepted from Pearson that the causes of any individual event widen out, quite unmanageably, into the history of the whole universe. So a deterministic treatment of single events is impossible.[72]

For Wright, the full significance of Pearson's statistical viewpoint only came in going way beyond anything Pearson sanctioned. For the second move was to monistic dual-aspect panpsychism. The hierarchy conclusion prompted by Moore's book contributed here, in suggesting that a molecule or atom might be like a minute organism, in that it has structure and incessant activity. So the properties of organisms seemed projectible all the way down the hierarchical series. What is more, when Wright considered, among other things, how evolutionary biology suggests the projection back into our remote animal ancestry of mind, there seemed no resolution of the problem of where mind came from originally, and that it was least arbitrary to have mind constituting the underlying reality of everything. Thus did Wright come to a panpsychist philosophy, one he was only later to learn resembled what Pearson's friend W. K. Clifford had embraced and, before him, Fechner and others, but one that Pearson himself never came close to endorsing.

It was the dual aspect feature of Wright's panpsychism — inner mind as reality, outer observable actions and appearances of mind as matter — that left science proceeding much as before, within precise metaphysical limitations. Introspective human psychology apart, it is not part of science to "make imaginative interpretations of the internal aspects of reality — what it is like, for example, to be a lion, an ant or an anthill, a liver cell, or a hydrogen ion". Science, accordingly, does not explain the behavior of an amoeba "as due partly to surface and other physical forces and partly to what the amoeba wants to do"; that would only lead to duplication, and so science sticks to the physical forces. In this program for science the "unique creative aspect of every event necessarily escapes" the scientist.[73]

On a statistical view of lawful causality, however, higher organisms, with their very numerous component parts, should "simulate complete determinism" in their actions. The reason they do not is that it is of the "essence of an organism" to contain many "switch or trigger mechanisms which bypass purely statistical behavior". Accordingly, at least since 1927 and Lindbergh's transatlantic flight, Wright illustrated this view by arguing that the direction of an aircraft's flight could be "fully accounted for deterministically except for the product of a succession of infinitesimals", that are traceable to the pilot's brain and are decisive because of the switch and trigger mechanisms, physical mechanisms, there and in the rest of his body and in the aircraft. "A high degree of freedom of choice by the whole is thus consistent with apparent deterministic behavior of the parts".[74]

It was after invoking this thesis in 1931 that Wright wrote that this view "implies considerable limitations on the synthetic phases of science", but "quantum physics" had shown, in any case, that "prediction can be expressed only in terms of probabilities, decreasing with the period of time". What then of evolution?

As to evolution, its entities, species and ecologic systems, are much less closely knit than individual organisms, One may conceive of the process as involving freedom, most readily traceable in the factor called here individual adaptability. This, however, is a subjective interpretation and can have no place in the objective scientific analysis of the problem.[75]

Forty five years later Wright reaffirmed this conclusion adding only that he would now write "individual selection" and "selective diffusion" in place of "individual adaptability", so as to refer to the "coefficients which are actually used in the mathematical formulation, but which nevertheless represent processes which may involve choices made by individual organisms".[76]

So, although the subjective viewpoint of "choices" does not belong in the scientific analysis, the physical bases of their outward manifestations — the trigger and switch mechanisms in any organisms — very much do. We have, here, therefore, a real locus for the influence of Wright's philosophy upon his biology. For he has made a judgment that, above the level of the individual organism, there is no organisation sufficiently integrated, no physical system that is closely knit enough, for any bypassing of statistical behavior such as trigger and switch mechanisms allow in any individual organism. Species, even ecologic systems, may be loosely called organisms, but they are too loosely organised to be judged organisms in the full sense that individual organisms should be.[77]

That this judgment was influential in Wright's science is apparent when we notice the central place taken by individual adaptability in his analysis of the evolutionary process in relation to genetics, including developmental no less than transmission genetics. "Individual adaptability is, in fact, distinctly a factor of evolutionary poise". It is not only of "the greatest significance as a factor of evolution in damping the effects of selection" and keeping these from being excessive compared to the inverse of the population size and to gene mutation rates, "it is itself perhaps the chief object of selection", he says, continuing:

The evolution of complex organisms rests on the attainment of gene combinations which determine a varied repertoire of adaptive cell responses in relation to external conditions. The older writers on evolution were often staggered by the seeming necessity of accounting for the evolution of fine details, of an adaptive nature, for example, the fine structure of all the bones. From the view that structure is never inherited as such, but merely types of adaptive cell behavior which lead to particular structures under particular conditions, the difficulty to a considerable extent disappears. The present difficulty is rather in tracing the inheritance of highly localized structural details to the more immediate inheritance of certain types of cell behavior.[78]

This is a line of thought that has no equivalent in Fisher. When Fisher addressed the worries of the older writers, he insisted that they were all taken care of by combining two resources: the resolution of those worries offered by Darwin in the *Origin*, and the resolution, made possible by heredity turning out to be particulate not blending, of any remaining difficulties that Darwin had had.[79] It is not that Wright's individualism — his emphasis on individual adaptability — traces solely to his preoccupation with the hierarchy of organisms from ions to biota, this tracing in turn to his philosophical debts to Moore; it is rather than he had philosophical as well as scientific, especially physiological-genetical, reasons for that emphasis.

It is easier, when we take these reasons into account, to see why Wright is an organismic as well as a genic selectionist, but not a group selectionist. What Wright calls intergroup selection is — remember the bulls — a selection as to which groups within the species will alter the whole species by exporting to the rest of the species successful genotypic combinations. That Wright has no level of selection higher than this is because he finds no sufficiently persistent and integrated entities above the level of individuals in the hierarchy of organisms. Conversely, his requiring of a theory of evolution that it bring out the consequences of persistence and variation at many levels of organisation, from genes up to individuals, reflects his finding that below the species there are several consequential levels of organismic integration. His inclination to make such requirements belongs to his philosophical life no less than to his scientific life, because the causal consequences of integrated organisation formed a subject that was common to his private philosophical quest and his public scientific career.

15. FISHER AND WRIGHT: COMPARISONS AND CONTRASTS

Sometimes, in a comparative and contrastive study, one can draw on those familiar distinctions that have often served well on other occasions: the distinctions made, to cite only a few examples, between classical and romantic outlooks; or tenderminded and toughminded philosophies; or this-worldly and other-worldly thinkers; or conservative and rationalist attitudes. Certainly those distinctions are relevant and fruitful when brought to the case of Fisher and Wright. But it will be evident already that no one of them fits well enough to be usefully developed in a sustained way.

Equally, the distinctions often brought to the analysis of the history and philosophy of biology do not provide obviously accurate bases for the comparisons and contrasts one needs to make between Fisher and Wright: reductionism versus holism, mechanism versus vitalism, and so on. To be sure, it is true that much in Fisher does seem to assimilate his science to the kind of atomistic, reductionist, determinist, mechanistic physics that is traditionally associated with Laplace and the Laplacean school two centuries ago now. Garland Allen has accordingly taken Fisher to be a specimen mechanist materialist — working in a billiard ball universe or, at least, with analogies drawn from such a universe — to be contrasted with the evolutionary geneticist I. M. Lerner as a specimen holist materialist.[80] Likewise, Jean Gayon, in his comprehensive and authoritative treatise on the history of the theory of natural selection from Darwin to Kimura, contrasts Fisher's Fundamental Theorem with Wright's adaptive topography, seeing the first as Laplacean in character and inspiration, while the other draws on a hierarchical conception of reality.[81] Now, the view taken of Wright in the present paper agrees almost entirely with Gayon's presentation, and indeed is indebted to it; the only serious point of divergence being a strategic one; where Gayon concentrates on the adaptive topography in Wright, the focus here has been on the theory of heterogenetic and homogenetic balance. However, where Gayon finds Fisher's statistical mechanical analogies aligning Fisher with Laplace, they have been seen, here, following Fisher himself, as distancing Fisher from that program in physics and as making him post-Boltzmannian rather than post-Laplacean in his indeterminism.

    But does there not remain a sense in which Fisher's science is, at least, more reductionist, less holistic than Wright's. Perhaps there does. However, rather than trying to identify that sense, it may be more valuable here to suggest that there is a decisive sense in which both Fisher and Wright are reductionistic in their theorising, but that Wright's reductions are much more complicated than Fisher's. Both are reductionist in one respect, surely. For both, the causal theory of evolution is to be reduced to Mendelian genetics through a statistical treatment of the consequences of that Mendelian genetics. The difference is that Wright's notions about the reducing science — Mendelian genetics — are more comprehensive and introduce more complications, most obviously physiological and developmental complications. Not coincidentally, his statistics takes him less of the way toward his evolutionary conclusions.

    It is implausible to say, therefore, that Fisher did things differently from Wright because he was more reductionist; by itself that judgment would do very little for us in our efforts to understand why their evolutionary theories came out so divergently. We have, instead, to characterise their science less abstractly; saying, perhaps, that their integrations of Darwinism, Mendelism and biometry differed as they did because, among other things, one combined that integrative ambition with eugenics, Darwin family connections and statistical mechanics, while the other came to it through Castle, physiological genetics and livestock breeding.

If we turn to characterise the two men's universes rather than their science, there are difficulties, again, in ensuring that we are comparing what is comparable. For Fisher, the relationship between philosophy and science was such that he sought a unified account of the physical and social world that grounded his philosophical arguments in scientific foundations. For Wright, there is no such movement from science to philosophy to give a single synthesis. This difference between the two is reflected in Wright's response to Fisher's thermodynamic analogies. Writing in 1931, in the *Journal of the American Statistical Association* on 'Statistical Theory of Evolution', Wright opened appropriately by summarising the "interesting comparison" made by Fisher on "the position of the evolutionary principle in biology" — significantly Wright uses the word evolution here, throughout, where Fisher would have talked of natural selection — and the second law of thermodynamics in physics. Having given Fisher's positive analogies, Wright gives only one of Fisher's disanalogies: the second law entails disorganisation, "a passage from less probable to more probable states"; evolution, as usually described, involves, he says, the very reverse. Tellingly Wright does not signal any direct disagreement with any of these Fisherian analogies or disanalogies. Instead, in keeping with his own view of the relationship between science and philosophy, he says that the following three options are "philosophical questions that I shall not attempt to discuss": whether "evolution is a mere eddy" in the general running down of the universe, or whether "the developmental side of nature" so conspicuous in biology is an "aspect of reality more basic than increase in entropy in physical systems" or whether "time is essentially" directionless. The Fisherian comparison, he concludes, "brings out the difficulty of accounting for the evolutionary process on the same basis, statistical theory", as that leading to the entropy law in physics. Yet, says Wright, switching back from philosophy to science: "it seems the only course open to scientific analysis"; for, as he explains, the possibility of a "physiological rather than statistical interpretation" is not available with the demise of Lamarckian inheritance.[82]

That Wright never accepted that evolution was a counterentropic

eddy in an otherwise entropic universe is only hinted at here or elsewhere. But years later he would end his piece on 'Organic Evolution'
for the *Encyclopedia Britannica* by saying that, although each species
has been "treated as if it evolved independently of the rest of the world
except for a rather mechanical-seeming connection through the concept
of selection pressure", in fact the "evolution of each species is merely an
aspect in the evolving pattern of life as a whole and indeed of the world
as a whole". Natural selection, he declares in closing, "is an abstraction
of the complicated reciprocal process" whereby the "pressure of the
species" to expand in the world and the pressure of the world "to keep
the species in its place" results, through devious ways, "in a general
trend toward progressive elaboration of the patterns of organization of
both".[83]

Where Fisher restricted his optimism to the future of civilisation,
provided intelligence was now allowed to take it under eugenic direction, Wright seems to have had a far wider cosmic hope for progress.
However, if that tempts us to think that more of traditional religion
lived on in Wright than in Fisher, we are checked by two reflections.
The first is that the nearest equivalent for Wright of Fisher's college
chapel was the Unitarian Church in Madison that Wright did indeed
attend; but it is no full equivalent, being a far less demanding institution, metaphysically speaking, and one where Wright never went so far
as to preach, it seems. The second is that when Wright did consider the
metaphysics of theism the outcome was negative. For anything physical,
such as a physical universe as a whole, to have a mind of its own, as a
person has, it must, argued Wright, have a comparably tightly knit
organisation; but this the universe, even with its ubiquitous fields of
force, gravitational and otherwise, seems manifestly to lack. When it
came to religion in general, Wright was no more a disbeliever than
Fisher, but where Fisher had his Christianity to reconcile with his
biology, Wright had only, it seems, some minimal Unitarian sympathies
to preserve.[84]

When we turn to the political elements in the case, we have a
procedural drawback to face. Fisher wore his politics on his sleeve,
most overtly when writing as one eugenist for others. By contrast
Wright never openly declared his political loyalties in print. There is,
however, an indirect way of getting at a rough, intuitive comparison and
contrast between the politics, in the broadest sense, of the two men.
This is to look at their stand on those topics in genetics that are most

ideologically telling: most especially, the nature-nurture controversy regarding human intelligence; human racial differences; the controversy over genetic damage from radiation; and eugenics itself.

Before doing this, however, we need to appreciate the difficulties in doing so. To some extent, it is possible to introduce a convenient simplifying assumption, by taking Fisher, as he is naturally viewed from a radical, socialist perspective; that is, as a pretty pure specimen of that conservative liberal ideology so manifestly adapted to the aspirations of the professional middle class in Edwardian England, as those aspirations were pursued through the eugenics movement. Fisher, as an extreme hereditarian and ultra-selectionist fits, almost perfectly, then, the historiographies that see Malthusianism and Social Darwinism as succeeded in the twentieth century by those other varieties of biological determinism that Fisher upholds. Nor shall we resist this assumption here. Perhaps the English eugenics movement was not so dominated by professional people among the middle class as was once thought, but for Fisher himself it was a movement especially appropriate to that social role.[85] Certainly Fisher rejects determinism in some senses, but he remains a biological determinist in holding that the properties in people decisive for the future prospects of civilisation are causally conditioned, and in that sense determined biologically, albeit probabilistically, in that their fate is only controllable for the best if the causation that biologists study — the causation of heredity and selection — is intelligently directed.

It is defensible, therefore, to ask how far Wright's thinking diverges from Fisher's in ways that might make him a less appropriate target for a radical, socialist critique of his thought. The answer, however, turns out once again to be less easy than one might think in advance.

On IQ and its heritability, Wright appears far too uncritical for radical, socialist comfort. Wright had taken up this topic, first, in a paper of 1931 and he returned to it at length in 1978 in the fourth volume of *Evolution and the Genetics of Populations*. He was certainly wary of drawing genetic comparisons among populations with different cultures; and he was certainly wary of saying how far IQ tests measured intelligence as that concept was generally understood. However, he saw these two cautions as consistent with taking IQ as a repeatable measure of some aspect of mental capacity, and with using data from studies of adopted children and from twin pairs to estimate its heritability within the population in question, and with siding more with Burt and Jensen than with their critics. Again, under the heading 'Temperament,' Wright joined those who had calculated a moderately strong heritability for such traits as responsibility, self-control and intellectual efficiency.[86]

One main conclusion from the following chapter, on 'Racial Differentiation in Mankind,' is that human evolution in the prehistoric and historic periods has consisted mainly of the "last phase of the shifting balance process": namely, "excess gene flow from a limited number of primary and secondary centers in which culture, and presumably genetic capabilities for it, have reached selective peaks". In working toward this conclusion, Wright avoids any assumption that there are genetic bases for this or that particular cultural trait. Earlier in that volume he had concluded that Fisherian individual selection and Hamiltonian "familial selection" are not entirely explanatorily adequate, and that "intergroup selection (Wynne-Edwards)" is also needed in explaining the evolution of "behavior". Now, he considers the "considerable but incomplete correlation" between the evolution of the genetic system and the evolution of culture. Stressing that if "the multiple genetic aspects of mental ability could be measured more independently of culture than is the case", then doubtless "each local race" would turn out to have "its own unique combination of favorable qualities", he urged also, however, that there have been "wide differences" among peoples "in average intellectual ability and cultural level from the standpoint of progress towards the situation in civilized man". The "capacity to anticipate and plan for the future" would be favored "under northern conditions and selected for insofar as it has a genetic basis".[87]

Such tentative conclusions and their lack of active policy implications would not have satisfied anyone who accepted Fisher's views on civilisation. When it came to radiation and genetics, Wright was, however, less reluctant to be normative. Wright's analysis rests on the assumption that "damage to society" can be assessed "more objectively than personal illness, pain, and frustration by classifying human phenotypes according to the ratio of social contribution to social cost", that is, cost of "rearing, education and maintaining standards of living". Thus, in an extreme case, a lethal mutation causing perinatal infant death is costly genetically, and distressing personally, but not very costly socially, because by contrast to a disabling but not fatal disease, there is no great burden to society. Not all "social damage" has genotypic bases; some is due to "reliance on inherited wealth", for instance; but there is more damage from those who "contribute negatively because of antisocial activities, and here genotype plays more role, though how much is a controversial question". Wright is hardly as unequivocal as the eugenists of old would have liked, but what he says is hardly grist for their opponents' mills either. Likewise, with the issues dividing Dobzhansky and Muller on the topic of genetic load. He distances himself from Muller, and sides silently with Dobzhansky, his onetime co-author, over the levels of heterozygosis it is proper to assume and over the advantages of those levels; but he concludes that Muller's campaign to reduce our exposure to radiation is "fully" justified.[88]

Muller was hardly less committed to eugenics than Fisher, but Wright kept his distance from their position. Although listed on its letterhead as one of 108 members of the advisory council of the American Eugenics Society, Wright never endorsed any eugenic policy. His reluctance stemmed, it seems, from the view that any positive eugenic measures required agreement as to what an ideal society would be, and no such agreement is achievable. "An industrial civilisation is a more complex organism" than an agricultural one and, therefore, he once wrote, it "requires development of many diverse types". He suspected that "the political and other choices" determining what sort of civilisation there is, tend "more or less automatically" to direct "the genetic character of the population", though perhaps with a considerable lag time. A "positive eugenic program" would then be directed at correcting the maladjustment, arising from this lag, either by changing the genetics to fit to social organisation, or by changing the social organisation "so as to fit types of individuals which seem admirable but have insufficient scope in the existing society". Needless to say, Fisher would never have agreed that an adjustment lag is all that needs correction, nor therefore that these corrective measures would be appropriate. But, although Wright's misgivings about anything like Fisher's eugenics are profound, they are insufficiently grounded in dissatisfaction with the existing social order to appeal to radical, socialist critics — or indeed, radical, conservative critics — of all eugenics.[89]

There is one impression that is hard to resist, but it does rather little for us historiographically. Fisher was rarely drawn to moderate, middle-of-the-road positions, while Wright often was; indeed his very language, as an evolutionary theorist, of balance and poise (meaning a very fine balancing) fits such a temperament all too obviously. Such impressions cannot take us, however, significantly beyond our earlier conclusions that, for instance, in evolutionary biology, we have in Fisher the most ardent, vindicative selectionist the world had ever — or perhaps has ever — seen, while in Wright we have an adjudicative theorist of evolution.

Quite generally, it does seem that the contrasts and comparisons one can draw between Fisher and Wright on political and ideological issues certainly contribute to the large picture, but more by way of confirming conclusions reached on other biographical grounds than by providing an ultimate resolution of anything that is otherwise inexplicable.

16. CONCLUSIONS

Returning to those themes announced at the opening of this paper, we may begin with the question of materialism, more precisely mechanistic materialism and its relations to Darwinian evolutionary theory. On the face of it, the case of Fisher and Wright must seem to make trouble for any view that Darwinian science has followed a succession from mechanistic to dialectical materialism. For consider where they stand, generationally speaking. Coming of age just before the first World War, they engaged a generation of earlier writers — such as Karl Pearson, Henri Bergson and Lloyd Morgan, to name only a few — who had refused to agree that evolution had to be interpreted mechanistically and materialistically. Fisher and Wright were to disagree with much that these authors proposed, either as science or as philosophy. However, each of the two men, in his different way, came to believe strongly that Darwinian biology does not require one to be mechanistic and materialistic in one's metaphysics.

Their cases suggest, therefore, that we have to be wary of constructing some historiographical abstraction — Darwinism, the Darwinian worldview or the Darwinian tradition or whatever — and then presuming that there is something that can be identified as the natural, dominant or inherent metaphysical character of this abstract construction. It may well be that Darwin himself — one thinks especially of his notebooks of the late 1830's — felt the need to make assumptions about consciousness and will that were deterministic (if not mechanistic) and materialistic, in order, as he saw it, to bring mind and so man within the causal, explanatory scheme he was developing for plant and animal structures and habits. It may well be that in assimilating Darwin's thinking some people, later, felt that they had to make similar assumptions. However, even to acknowledge this much is to leave open the possibility that yet others have felt differently, so that we will only be begging historical questions if we lay down an essentialist stipulation that insofar as these others did this then they were not really true Darwinians. Richards's history seems better fitted to avoid this kind of question begging than Allen's and is, therefore, confirmed to that extent at least. Quite generally, what we need to take in is the possibility that longstanding empirical convictions and metaphysical anxieties about mind and body, will and consciousness, creativity and causality and so on have continued to live on among professional twentieth-century biologists, rather than being left behind with the passing of the great nineteenth-century debates among men of science, men of letters and men of the cloth. The disagreements between Allen, Greene and Richards have done us a welcome favor in making us aware of that possibility.[90]

Turning now to the Bowler theses, one has to note that whether these are seen to be confirmed depends very much on how one interprets them. Sometimes, Bowler seems to be saying that it was idealistic philosophies of nature that fostered progressionist and developmentalist views of evolution, so that, conversely, the new Mendelian genetics of the twentieth century could be effective in discrediting such views because it challenged those philosophies of nature.[91] But this thesis is surely not supported by the case of Fisher and Wright. In Wright we have someone who saw his own philosophy as explicitly idealistic; and even Fisher can not be seen as confronting and repudiating idealism in the philosophy of nature. What does one need to concentrate on, instead, therefore, in understanding Fisher's and Wright's departures from the progressionist and developmentalist views so sidespread in the sixty years after Darwin's *Origin*? Here, surely, the obvious answer provides the right place to start, at least, and that is the conjunction in these two of Mendelism, biometry and Darwinism.

A further reflection is also pertinent. The biometricians, as led by Weldon and Pearson, had already upheld an interpretation of evolution that was both selectionist and nondevelopmentalist. One possible way to understand Fisher and Wright, is to see them as bringing Mendelism into conjunction with biometry in such a way that the synthesis retained this selectionism and lack of developmentalism. To follow up this interpretation, one would have to ask, further, how their Mendelism could be brought into such a synthesis in that way. Here, surely, it

would be relevant to cite those features of heritable Mendelian variation — as Fisher and Wright conceived of it — that made it, in and by
itself, quite incapable of directing the process of evolution: multiple
factors, each of small effect, influencing quantitative traits; the random
shuffling of factors in recombination without linkage; the rarity and
fortuitousness of mutations and so on.

Bowler himself would not deny that these things are directly relevant, but he would rather, it seems, rest his case on a more general
thesis: namely, that, with the new Mendelian conception of heredity, it
became possible to distinguish transmission genetics from developmental genetics, and so to divorce evolution from development by
integrating the theory of natural selection with Mendelian transmission
genetics. This separatism thesis works well, it must be conceded, for
Fisher. But it is hardly acceptable without qualification for Wright. It
goes without saying, naturally, that the contrasts between mitosis and
meiosis are fundamental for Wright's understanding of the contrast
between development and evolution. It also goes without saying that, as
Wright put it himself, "chance variation" and the "influence of the
environment" are both "denied in the Aristotelian conception of evolution as analogous to individual development, the realization of an
innate potentiality, irrespective of environment (except for provision of
the necessary conditions for life)". But it does not follow from these
familiar points that the problems of developmental genetics became
entirely separable from those of evolutionary genetics. Indeed, Wright
constantly relates to one another his views on these two subjects. Thus,
in explaining the differences between Richard Goldschmudt's macro-
mutationist views and his own and others' "neo-Darwinism", he stressed
that there lay at the root of those differences divergent conceptions of
the "physiological relations between germ plasm and organism".
Goldschmidt seems to hold, he says, that "the conception of the organism as an integrated reaction system requires a corresponding *spatial*
integration of the germ plasm" so that essential change in the reaction
system can only come about "by repatterning of the chromosomes". To
himself and others, Wright says, "a *temporal* integration is all that is
necessary, or even possible, with the chain reaction as the simplified
model". He goes on to explain how, with branching hierarchical systems
of chain reactions, there can be many reaction systems based on the
same set of genes and such systems can then evolve "more or less
independently of one another".[92]

In Wright's case, therefore, it seems best not to say that evolutionary biology is tied to transmission genetics while sundered from developmental genetics, but to say that the integration of a new Mendelian genetics, of transmission and development, with Darwinism and biometry, was wrought in such a way that developmentalist evolutionary theory was undercut and an opportunistic, arboriform process was seen to result, as by Darwin himself, from the contingencies of variation interacting with the contingencies of environmental influences. In so far as Wright is a representative figure, therefore, the emergence within the Mendelian program of a new physiological and developmental genetics, must be included within the history of evolutionary biology in the early twentieth century.

So many probabilistic themes have proven relevant to the case of Fisher and Wright that the only conclusion to be confidently drawn is that there is no single conclusion. Even if we focus on the departure of science from some supposed Laplacean determinism — that may or may not have been a genuine historical reality — the story remains highly complicated.[93] Among the many reasons for this complexity four can usefully be stressed here. First, as several authors have insisted lately, there is more, far more, to the history of statistical thinking than the history of mathematical statistics. Were one to concentrate on Fisher's and Wright's mathematical consensus one would miss almost all the interesting and instructive comparisons and contrasts to be made between their lives and works as evolutionary theorists. Second, as these same authors have urged, since the mid-nineteenth century statistical thinking has ranged over a remarkable array of domains from social theory to astronomy, and statistics has itself developed as it has because of that wide range of applications. Wright and Fisher, in their different ways, vividly instantiate that theme. Fisher learned the style, habits and skills of his statistical thinking about genetics and evolution from Jeans and Boltzmann on gases as much as from Yule and Pearson on parental and fraternal correlations; while for Wright a decisive early statistical ambition was to move beyond Jennings's analyses of systems of livestock breeding.

Third, the scientific implications of statistical thinking have never been straightforward in their interpretation. Pearson and others saw statistics, positivistically, as leading science away from causation to correlations and laws alone. Wright and Fisher agreed in taking precisely the opposite view; for them statistical techniques were a way to further the traditional aim of science in finding the causes, often hidden causes, behind phenomena. Indeterministic or probabilistic causation was never reducible to statistical correlation. Fourth, the philosophical implications of statistical thinking have been even less straightforward in their interpretation. By the time of Fisher's and Wright's undergraduate years it was a commonplace that, in physics itself, mechanics was, arguably at least, no longer as mechanistic as it had been, nor matter theory as materialistic, nor causation as deterministic; that, in sum, the Laplacean intellect was no longer an unproblematic ideal. But, equally, it was evident that there was no agreement as to how to understand the ultimate philosophical significance of these developments. What the serious but amateurish philosophising of our two young evolutionary theorists shows is that there was no one simple and persuasive response that it was obviously correct to make to that rich, rewarding but confusing state of discussion. As on other occasions, therefore, a virtue of history, especially when it includes biography, is that we can learn how, in the long run, things are never likely to go as simply as either our philosophical predelictions or our scientific upbringing may incline us to presume.[93]

University of Leeds

NOTES AND REFERENCES

[1] Almost everything in this paper is indebted to extensive discussions over the years with John Turner and to his papers on Fisher and Wright; so much so that large parts could appropriately be thought of as jointly authored. The paper also draws throughout on the publications and conversation of William Provine. I am grateful, too, for valuable talks with Garland Allen, John Beatty, James Crow, Anthony Edwards, Jonathan Harwood, Ernst Mayr, Robert Olby and David West.

[2] For a fuller discussion of these and other rationales, see the editors' introduction and other contributions to R. C. Olby, G. N. Cantor, J. R. R. Christie and M. J. S. Hodge, eds., *Companion to the History of Modern Science*, London and New York: Routledge, 1990.

[3] G. Allen, 'The Several Faces of Darwin: Materialism in Nineteenth and Twentieth-Century Evolutionary Theory', in D. S. Bendall, ed., *Evolution from Molecules to Men*, Cambridge: Cambridge University Press, 1983; pp. 81–102; J. C. Greene, *Science, Ideology and World View*, Berkeley: University of California Press, 1981; R. J. Richards, *Darwin and the Emergence of Evolutionary Theories of Mind and Behavior*, Chicago and London: The University of Chicago Press, 1987; pp. 405–407.

[4] P. J. Bowler, *The Non-Darwinian Revolution: Reinterpreting a Historical Myth*,

Baltimore: Johns Hopkins University Press, 1988; *The Mendelian Revolution: The Emergence of Hereditarian Concepts in Modern Science and Society*, London: Athlone Press, 1989; *Charles Darwin: The Man and His Influence*, Oxford and Cambridge, Mass.: Basil Blackwell, 1990.

[5] L. Krüger *et al.*, eds., *The Probabilistic Revolution*, 2 vols., Cambridge, Mass: MIT Press, 1987; T. M. Porter, *The Rise of Statistical Thinking, 1820—1900*, Princeton: Princeton University Press, 1986; S. Stigler, *The History of Statistics: The Measurement of Uncertainty before 1900*, Cambridge, Mass.: Harvard University Press, 1986; G. Gigerenzer *et al.*, *The Empire of Chance. How Probability Changed Science and Everyday Life*, Cambridge: Cambridge University Press, 1989; Ian Hacking, *The Taming of Chance*, Cambridge: Cambridge University Press, 1990.
[6] J. F. Box, *R. A. Fisher: The Life of a Scientist*, New York: John Wiley, 1978; J. H. Bennett, ed., *Natural Selection, Heredity and Eugenics. Including selected correspondence of R. A. Fisher with Leonard Darwin and others*, Oxford: Clarendon Press, 1983; J. H. Bennett, ed., *Statistical Inference and Analysis. Selected Correspondence of R. A. Fisher*, Clarendon Press, 1990; W. Provine, *Sewall Wright and Evolutionary Biology*, Chicago: University of Chicago Press, 1986.
[7] Provine, *Wright*, pp. 18 and 95—7; Wright, 'Panpsychism and Science', in J. E. Cobb and D. R. Griffin, eds., *Mind in Nature*, Washington: University Press of America, 1975, pp. 79—88; J. F. Crow, 'Sewall Wright's Place in Twentieth-Century Biology', *Journal of the History of Biology* 23 (1990) 57—89. Further evidence of Wright's philosophical views, see items 26, 38 and 104 in the full bibliography in Provine, *Wright* and also in *Evolution. Selected Papers* (see n. 14).
[8] Box, *Fisher*, pp. 288—95; *The Social Selection of Human Fertility*, Oxford: Clarendon Press, 1932; 'Indeterminism and Natural Selection', *Philosophy of Science* 1 (1934) 99—117; *Creative Aspects of Natural Law*, Cambridge: Cambridge University Press, 1950. All three pieces are reprinted in facsimile in J. H. Bennett, ed., *Collected Papers of R. A. Fisher*, 5 vols., Adelaide: The University of Adelaide, 1971—4. In citations of Fisher's writings that are in this invaluable collection, I shall refer to them by volume number together with the number of the item given it by Bennett. These three are, accordingly, at *CP*, *3*: 99 and 121 and *5*: 241. Box and the first volume of *CP* contain a full bibliography.
[9] Fisher, *Eugenics Review* 23 (1932) 88—90, in Bennett, ed., *Natural Selection*, pp. 287—8.
[10] Fisher, 'On the Dominance Ratio', *Proceedings of the Royal Society of Edinburgh* 42 (1922) 321—41; *CP*, *1*: 24.
[11] Provine, *Wright*, pp. 232—276; *Origins of Theoretical Population Genetics*, Chicago: University of Chicago Press, 1971.
[12] Fisher, 'Darwinian Evolution of Mutations', *Eugenics Review* 14 (1922) 31—34, pp. 33—4; *CP*, *1*: 26.
[13] Provine, *Wright*, pp. 232—276.
[14] Wright, 'The Genetical Theory of Natural Selection. A Review', *Journal of Heredity* 21 (1930) 349—59, pp. 352 and 355. Where, as in this case, the item is reprinted in facsimile in S. Wright, *Evolution. Selected Papers. Edited and with Introductory Materials by W. B. Provine*, Chicago: University of Chicago Press, 1988, I shall give the item number and the page references there (this collection does not reproduce the

original page numbers, although of course they can be reconstructed from what is given). Thus, in this case, *ESP, 9*: 83 and 86.

[15] In addition to Box, Bennett and Provine as already cited, see Provine, 'The R. A. Fisher—Sewall Wright Controversy and Its Influence upon Modern Evolutionary Biology', *Oxford Surveys in Evolutionary Biology* **2** (1985) 197—200, and J. R. G. Turner, 'Random Genetic Drift, R. A. Fisher, and the Oxford School of Ecological Genetics', in L. Krüger *et al.*, eds., *The Probabilistic Revolution* **2**, pp. 313—54.

[16] In addition to Box and Bennett, see B. J. Norton, 'Metaphysics and Population Genetics: Karl Pearson and the Background to Fisher's Multifactorial Theory of Inheritance', *Annals of Science* **32** (1975) 537—53; 'Fisher and the Neo-Darwinian Synthesis', in E. G. Forbes, ed., *The Human Implications of Scientific Advance. Proceedings of the 15th International Congress on the History of Science*, Edinburgh: Edinburgh University Press, 1978; 'Fisher's Entrance into Evolutionary Science: The Role of Eugenics', in M. Grene, ed., *Dimensions of Darwinosm*, Cambridge: Cambridge University Press, 1983, pp. 19—29; 'La Situation Intellectuelle au Moment des Débuts de Fisher en Génétique des Populations', *Revue de Synthése. IIIe serie: 103—4* (1981) 230—250. (A volume devoted to *R. A. Fisher et l'Histoire de la Génétique des Populations*).

[17] On mutations and their rates, see R. C. Olby, 'La théorie génétique de la sélection naturelle vue par un historien', *Revue de Synthése, IIIe serie: 103—4* (1981) 251—289. See the comments of Benett, *Natural Selection* p. 17, on the interpretation of Fisher given by various historians of science.

[18] The tradition of identifying natural selection as a known cause is discussed in M. J. S. Hodge, 'Natural Selection as a Causal, Empirical, and Probabilistic Theory', in L. Krüger *et al.*, eds., *The Probabilistic Revolution*, vol. 2, pp. 233—270 and 'Darwin's Theory and Darwin's Argument', in M. Ruse, ed., *What the Philosophy of Biology is. Essays dedicated to David Hull*, Dordrecht, Boston, London: Kluwer, 1989, pp. 163—182.

[19] Fisher, 'The Measurement of Selective Intensity', *Proceedings of the Royal Society of London, B* **121** (1936) 58—62, p. 58; *CP, 3*: 147.

[20] For a more extensive interpretation of Fisher's evolutionary theory along the lines taken here, see J. R. G. Turner, 'Fisher's Evolutionary Faith and the Challenge of Mimicry', *Oxford Surveys in Evolutionary Biology* **2** (1985) 159—196 and his *op.cit.* n. 15. For a comprehensive analysis of how Fisher's views stand today, see the two-part article, E. G. Leigh, 'Ronald Fisher and the Development of Evolutionary Theory', *Oxford Surveys in Evolutionary Biology* **3** (1986) 187—233 and **4** (1987) 212—263.

[21] See the articles of Turner, Olby and Leigh cited in notes 15, 17 and 20.

[22] Fisher, *The Genetical Theory of Natural Selection*, Oxford: Clarendon Press, 1930, and the second edition, published by Dover, New York, 1958.

[23] 'Paper on "Heredity" (comparing methods of Biometry and Mendelism)', in Bennett, ed., *Natural Selection*, pp. 51—8; the quoted words are at pp. 56—7.

[24] R. A. Fisher and C. S. Stock, 'Cuenot on Preadaptation. A Criticism', *Eugenics Review* **7** (1915) 46—61; p. 60. *CP, 1*: 5.

[25] *Ibid.*, pp. 60—1.

[26] 'On the Dominance Ratio' (see n. 10), p. 321.

[27] *Ibid.*, pp. 321—2. Provine, *Wright*, p. 241, has Fisher here making the investigation of selection comparable to the theory of gases. But this is surely an incorrect reading.

[28] *Ibid.*, p. 340; 'Croonian Lecture: Population Genetics', *Proceedings of the Royal Society of London, B* **141** (1953) 510—523; p. 515. *CP, 5:* 252.

[29] *Genetical Theory,* 1st ed., p. 11; 2nd ed., p. 11.

[30] *Ibid.*

[31] *Ibid.*, pp. 35 and 46 (1st ed.); pp. 37 and 50 (2nd ed.).

[32] G. R. Price, 'Fisher's "Fundamental Theorem" Made Clear', *Annals of Human Genetics, London* **36** (1972) 129—140, pp. 132 and 131. Price argues (p. 131) that since "the standard of 'fitness' changes from instant to instant, this constant improving tendency of natural selection does not necessarily get anywhere in terms of increasing 'fitness' as measured by any fixed standard," and mean fitness is "about as likely to decrease under selection as to increase".

[33] See Wright's review cited in n. 14.

[34] 1st ed., pp. 35—37; 2nd ed., pp. 38—40.

[35] 1st ed., pp. 36—7; 2nd ed., p. 39.

[36] 1st ed., p. 37; 2nd ed., pp. 39—40. Fisher seems never to have gone explicitly in quest of "some more general principle".

[37] 'The Social Selection of Human Fertility' (see n. 8), pp. 8—9.

[38] 'Retrospect of the Criticisms of the Theory of Natural Selection', in J. S. Huxley *et al.*, eds., *Evolution as a Process*, London: Allen and Unwin, 1953, pp. 84—98. *CP, 5:* 314. This article was evidently written in the early 1930's. See Bennett, ed., *Natural Selection*, p. 202.

[39] In 1934, in 'Indeterminism and Natural Selection' (n. 8), Fisher wrote (pp. 110—111) revealingly: "Two characteristics of Darwin's evolutionary thought which, in the writer's opinion, give it its supreme and lasting value are, first, that while an active theorist, willing to follow long chains of reasoning if he felt their foundations secure, he constantly brought his speculations into contact with a candid and thoughtful scrutiny of living things themselves; and, secondly, that he was never willing to curb or limit his thought by theoretical considerations brought in from other fields of study. Thus his theory of evolution made demands upon the age of the earth beyond what the physicists, for many years after his death, regarded as physically possible. We now know that the evolutionists were right, and that it was the physical theory that was faulty. Again, in regard to men and his mental and moral characteristics, (his Soul in all but the meaning given to the word by theological theory), Darwin did not shrink from the plain meaning of the facts that fighting cocks could be bred for courage, or dogs for fidelity, and that the amazing plumage and ornaments of many male birds indicated a discriminative choice on the part of the females, which only an anthropocentric prejudice would hesitate to call aesthetic appreciation."

[40] 'Student', *Annals of Eugenics* **9** (1939) 1—9. *CP, 4:* 165.

[41] 'Paper on "Evolution and Society" ', in Bennett, ed., *Natural Selection*, pp. 58—62. See pp. 58, 61 and 62.

[42] 'The Evolution of Sexual Preference', *Eugenics Review* **7** (1915) 184—192. See pp. 191—2. *CP, 1:* 6. On Fisher's Nietzschean friends, see Box, *Fisher*, p. 18.

[43] 'Indeterminism and Natural Selection' (n. 8) p. 99.

[44] *Ibid.*

[45] *The Social Selection of Fertility* (n. 8).

[46] 'Indeterminism and Natural Selection' and *The Social Selection of Fertility.*

[47] *Creative Aspects of Natural Law* (n. 8), pp. 4—5.

[48] *Ibid.*

[49] *Ibid.,* p. 7.

[50] *Ibid.* p. 13. Articles on Bergson and on Smuts by T. A. Goudge (in P. Edwards, *Encyclopedia of Philosophy,* 8 vols., New York and London, MacMillan, 1967) give good introductions to their evolutionary views.

[51] *Ibid.,* pp. 20—22. See also 'The Renaissance of Darwinism', *Listener* 37 (1947) 1001. *CP, 4:* 217, and John Turner's two articles (nn. 15 and 20).

[52] Fisher to C. S. Sherrington: 22 January 1947 in Bennett, ed., *Natural Selection,* pp. 261—2.

[53] Here, as always in discussing Wright's career, I am drawing directly on Provine, *Wright.*

[54] *Ibid.,* pp. 180—1 and p. 16.

[55] See the three Wright articles cited in n. 7, and Provine, *Wright,* p. 95.

[56] 'The Relation of Livestock Breeding to Theories of Evolution', *Journal of Animal Science* 46 (1978) 1192—1200. *ESP, 1.*

[57] J. F. Crow, 'Sewall Wright's Place in Twentieth-Century Biology', *Journal of the History of Biology* 23 (1990) 57—89. Cf. Provine, *Wright,* p. 235.

[58] *Ibid.*

[59] 'Evolution in Mendelian Populations', *Genetics* 16 (1931) 97—159; p. 142. *ESP, 11:* 143.

[60] 'The Shifting Balance Theory and Macroevolution', *Annual Review of Genetics* 16 (1982) 1—19, pp. 4—6.

[61] 'The Roles of Mutation, Inbreeding, Crossbreeding and Selection in Evolution', *Proceedings of the Sixth International Congress of Genetics* 1 (1982) 256—66. *ESP, 12:* 161—171.

[62] 'Evolution', pp. 142—3, 100 and 98.

[63] *Ibid.,* pp. 144—5.

[64] *Ibid.,* p. 145.

[65] *Ibid.,* pp. 145—6.

[66] *Ibid.,* p. 150.

[67] *Ibid.,* p. 151.

[68] *Ibid.,* pp.157—8.

[69] 'Gene and Organism', *American Naturalist* 87 (1953) 5—18; 'Biology and the Philosophy of Science', *Monist* 48 (1964) 265—90. 'Panpsychism and Science' (n. 7).

[70] 'Panpsychism and Science', p. 80.

[71] 'Biology and the Philosophy of Science', p. 281.

[72] *Ibid.*

[73] 'Gene and Organism", p. 18.

[74] 'Panpsychism and Science', p. 84.

[75] 'Evolution', p. 154.

[76] 'Panpsychism and Science', p. 87.

[77] 'Evolution', pp. 154—5. I see my analysis here as consistent with the rather different one given in Provine, *Wright,* pp. 95—7.

[78] *Ibid.,* p. 147

[79] See Fisher, 'Retrospect' (n. 38).

[80] See Allen's paper cited in n. 3.

[81] J. Gayon, *La Théorie de la Selection: Darwin et l'aprés Darwin*. Thése de Doctorat, Université de Paris I, 2 vols., 1989; vol. 2, pp. 646—7. This important study is expected to appear in a revised, published version in 1992.

[82] 'Statistical Theory of Evolution', *Journal of the American Statistical Association* **26** (1931) supplement: 201—8; p. 201. *ESP 10*: 89—96.

[83] 'Evolution, Organic', in *Encyclopedia Britannica*, 14th ed. revised, 1948, vol. 8, pp. 915—29; p. 929. *ESP 26*: 524—538.

[84] Provine, *Wright*, p. 460; 'Panpsychism and Science', p. 85.

[85] On the ideology and politics of English eugenics see D. A. Mackenzie, *Statistics in Britain 1865—1930: The Social Construction of Scientific Knowledge*, Edinburgh: Edinburgh, 1981; R. C. Olby, 'The Dimensions of Scientific Controversy: The Biometrician-Mendelian Debate', *British Journal for the History of Science* **22** (1989) 299—320; G. R. Searle, 'Eugenics and Class', in C. Webster, (ed.) *Biology, Medicine and Society 1840—1940*, Cambridge: Cambridge University Press, 1981, pp. 217—242; D. Kevles, *In the Name of Eugenics*, New York: Knopf, 1985: E. J. Larson, 'The Rhetoric of Eugenics: Expert Authority and the Mental Deficiency Bill', *British Journal for the History of Science* **24** (1991) 45—60. For a radical, socialist perspective, see R. Lewontin, S. Rose and L. Kamin, *Not in our Genes. Biology, Ideology and Human Nature*, New York: Pantheon, 1985. It is their notion of biological determinism that I am referring to here.

[86] 'Statistical Methods in Biology', *Journal of the American Statistical Association* **26** (1931) supplement: 155—163; *Evolution and the Genetics of Populations*, 4 vols., Chicago and London: University of Chicago Press, 1968—1978, vol. 4, pp. 296—420.

[87] *Ibid.*, pp. 53, 458 and 455—6.

[88] *Ibid.*, vol. 3, pp. 477—9. Wright's fullest treatment of these issues is in 'Discussion on Population Genetics and Radiation', *Journal of Cell and Comparative Physiology* **35** (1950) 187—210 and 'On the Appraisal of Genetic Effects of Radiation in Man', in *The Biological Effects of Atomic Radiation: Summary Reports from a Study by the National Academy of Sciences*, Washington: NAS-NRC, 1960, pp. 18—24.

[89] Provine, *Wright*, p. 180 and Wright to P. Popenoe, May 13, 1931, as printed there on p. 181.

[90] See n. 3.

[91] See n. 4.

[92] 'Evolution, Organic' (n. 83), p 924; 'The Material Basis of Evolution', *Scientific Monthly* **53** (1941) 165—70; pp. 168—9. *ESP, 27*: 389—94.

[93] (Added in proof). Jean Gayon's book (see n. 81) now interprets Fisher much as I have here: *Darwin et l'après-Darwin*, Paris: Editions Kimé, 1992. On Philosophy and the architects of the synthetic theory, see also J. Harwood, 'Metaphysical Foundations of the Evolutionary Synthesis', forthcoming.

INDEX

Boston Studies in the Philosophy of Science

45. A. Ishimoto (ed.): Japanese Studies in the History and Philosophy of Science. (In prep.)   ISBN 90-277-0733-3
46. P.L. Kapitza: *Experiment, Theory, Practice*. Articles and Addresses. Edited by R.S. Cohen. 1980   ISBN 90-277-1061-9; Pb 90-277-1062-7
47. M.L. Dalla Chiara (ed.): *Italian Studies in the Philosophy of Science*. 1981
                                                ISBN 90-277-0735-9; Pb 90-277-1073-2
48. M.W. Wartofsky: *Models*. Representation and the Scientific Understanding. [Synthese Library 129] 1979   ISBN 90-277-0736-7; Pb 90-277-0947-5
49. T.D. Thao: *Phenomenology and Dialectical Materialism*. Edited by R.S. Cohen. 1986   ISBN 90-277-0737-5
50. Y. Fried and J. Agassi: *Paranoia*. A Study in Diagnosis. [Synthese Library 102] 1976   ISBN 90-277-0704-9; Pb 90-277-0705-7
51. K.H. Wolff: *Surrender and Cath*. Experience and Inquiry Today. [Synthese Library 105] 1976   ISBN 90-277-0758-8; Pb 90-277-0765-0
52. K. Kosík: *Dialectics of the Concrete*. A Study on Problems of Man and World. 1976   ISBN 90-277-0761-8; Pb 90-277-0764-2
53. N. Goodman: *The Structure of Appearance*. [Synthese Library 107] 1977
                                                ISBN 90-277-0773-1; Pb 90-277-0774-X
54. H.A. Simon: *Models of Discovery* and Other Topics in the Methods of Science. [Synthese Library 114] 1977   ISBN 90-277-0812-6; Pb 90-277-0858-4
55. M. Lazerowitz: *The Language of Philosophy*. Freud and Wittgenstein. [Synthese Library 117] 1977   ISBN 90-277-0826-6; Pb 90-277-0862-2
56. T. Nickles (ed.): *Scientific Discovery, Logic, and Rationality*. 1980
                                                ISBN 90-277-1069-4; Pb 90-277-1070-8
57. J. Margolis: *Persons and Mind*. The Prospects of Nonreductive Materialism. [Synthese Library 121] 1978   ISBN 90-277-0854-1; Pb 90-277-0863-0
58. G. Radnitzky and G. Andersson (eds.): *Progress and Rationality in Science*. [Synthese Library 125] 1978   ISBN 90-277-0921-1; Pb 90-277-0922-X
59. G. Radnitzky and G. Andersson (eds.): *The Structure and Development of Science*. [Synthese Library 136] 1979
                                                ISBN 90-277-0994-7; Pb 90-277-0995-5
60. T. Nickles (ed.): *Scientific Discovery*. Case Studies. 1980
                                                ISBN 90-277-1092-9; Pb 90-277-1093-7
61. M.A. Finocchiaro: *Galileo and the Art of Reasoning*. Rhetorical Foundation of Logic and Scientific Method. 1980   ISBN 90-277-1094-5; Pb 90-277-1095-3
62. W.A. Wallace: *Prelude to Galileo*. Essays on Medieval and 16th-Century Sources of Galileo's Thought. 1981   ISBN 90-277-1215-8; Pb 90-277-1216-6
63. F. Rapp: *Analytical Philosophy of Technology*. Translated from German. 1981
                                                ISBN 90-277-1221-2; Pb 90-277-1222-0
64. R.S. Cohen and M.W. Wartofsky (eds.): *Hegel and the Sciences*. 1984
                                                            ISBN 90-277-0726-X
65. J. Agassi: *Science and Society*. Studies in the Sociology of Science. 1981
                                                ISBN 90-277-1244-1; Pb 90-277-1245-X

Boston Studies in the Philosophy of Science

66. L. Tondl: *Problems of Semantics*. A Contribution to the Analysis of the Language of Science. Translated from Czech. 1981
ISBN 90-277-0148-2; Pb 90-277-0316-7

67. J. Agassi and R.S. Cohen (eds.): *Scientific Philosophy Today*. Essays in Honor of Mario Bunge. 1982   ISBN 90-277-1262-X; Pb 90-277-1263-8

68. W. Krajewski (ed.): *Polish Essays in the Philosophy of the Natural Sciences*. Translated from Polish and edited by R.S. Cohen and C.R. Fawcett. 1982
ISBN 90-277-1286-7; Pb 90-277-1287-5

69. J.H. Fetzer: *Scientific Knowledge*. Causation, Explanation and Corroboration. 1981   ISBN 90-277-1335-9; Pb 90-277-1336-7

70. S. Grossberg: *Studies of Mind and Brain*. Neural Principles of Learning, Perception, Development, Cognition, and Motor Control. 1982
ISBN 90-277-1359-6; Pb 90-277-1360-X

71. R.S. Cohen and M.W. Wartofsky (eds.): *Epistemology, Methodology, and the Social Sciences*. 1983.   ISBN 90-277-1454-1

72. K. Berka: *Measurement*. Its Concepts, Theories and Problems. Translated from Czech. 1983   ISBN 90-277-1416-9

73. G.L. Pandit: *The Structure and Growth of Scientific Knowledge*. A Study in the Methodology of Epistemic Appraisal. 1983   ISBN 90-277-1434-7

74. A.A. Zinov'ev: *Logical Physics*. Translated from Russian. Edited by R.S. Cohen. 1983   ISBN 90-277-0734-0
See also Volume 9.

75. G-G. Granger: *Formal Thought and the Sciences of Man*. Translated from French. With and Introduction by A. Rosenberg. 1983   ISBN 90-277-1524-6

76. R.S. Cohen and L. Laudan (eds.): *Physics, Philosophy and Psychoanalysis*. Essays in Honor of Adolf Grünbaum. 1983   ISBN 90-277-1533-5

77. G. Böhme, W. van den Daele, R. Hohlfeld, W. Krohn and W. Schäfer: *Finalization in Science*. The Social Orientation of Scientific Progress. Translated from German. Edited by W. Schäfer. 1983   ISBN 90-277-1549-1

78. D. Shapere: *Reason and the Search for Knowledge*. Investigations in the Philosophy of Science. 1984   ISBN 90-277-1551-3; Pb 90-277-1641-2

79. G. Andersson (ed.): *Rationality in Science and Politics*. Translated from German. 1984   ISBN 90-277-1575-0; Pb 90-277-1953-5

80. P.T. Durbin and F. Rapp (eds.): *Philosophy and Technology*. [*Also* Philosophy and Technology Series, Vol. 1] 1983   ISBN 90-277-1576-9

81. M. Marković: *Dialectical Theory of Meaning*. Translated from Serbo-Croat. 1984   ISBN 90-277-1596-3

82. R.S. Cohen and M.W. Wartofsky (eds.): *Physical Sciences and History of Physics*. 1984.   ISBN 90-277-1615-3

83. É. Meyerson: *The Relativistic Deduction*. Epistemological Implications of the Theory of Relativity. Translated from French. With a Review by Albert Einstein and an Introduction by Milič Čapek. 1985   ISBN 90-277-1699-4

Boston Studies in the Philosophy of Science

Boston Studies in the Philosophy of Science

Boston Studies in the Philosophy of Science

121. P. Nicolacopoulos (ed.): *Greek Studies in the Philosophy and History of Science.* 1990   ISBN 0-7923-0717-8
122. R. Cooke and D. Costantini (eds.): *Statistics in Science.* The Foundations of Statistical Methods in Biology, Physics and Economics. 1990
                                                          ISBN 0-7923-0797-6
123. P. Duhem: *The Origins of Statics.* Translated from French by G.F. Leneaux, V.N. Vagliente and G.H. Wagner. With an Introduction by S.L. Jaki. 1991
                                                          ISBN 0-7923-0898-0
124. H. Kamerlingh Onnes: *Through Measurement to Knowledge.* The Selected Papers, 1853-1926. Edited and with an Introduction by K. Gavroglu and Y. Goudaroulis. 1991   ISBN 0-7923-0825-5
125. M. Čapek: *The New Aspects of Time: Its Continuity and Novelties.* Selected Papers in the Philosophy of Science. 1991   ISBN 0-7923-0911-1
126. S. Unguru (ed.): *Physics, Cosmology and Astronomy, 1300-1700.* Tension and Accommodation. 1991   ISBN 0-7923-1022-5
127. Z. Bechler: *Newton's Physics on the Conceptual Structure of the Scientific Revolution.* 1991   ISBN 0-7923-1054-3
128. É. Meyerson: *Explanation in the Sciences.* Translated from French by M-A. Siple and D.A. Siple. 1991   ISBN 0-7923-1129-9
129. A.I. Tauber (ed.): *Organism and the Origins of Self.* 1991
                                                          ISBN 0-7923-1185-X
130. F.J. Varela and J-P. Dupuy (eds.): *Understanding Origins.* Contemporary Views on the Origin of Life, Mind and Society. 1992   ISBN 0-7923-1251-1
131. G.L. Pandit: *Methodological Variance.* Essays in Epistemological Ontology and the Methodology of Science. 1991   ISBN 0-7923-1263-5
132. G. Munévar (ed.): *Beyond Reason.* Essays on the Philosophy of Paul Feyerabend. 1991   ISBN 0-7923-1272-4
133. T.E. Uebel (ed.): *Rediscovering the Forgotten Vienna Circle.* Austrian Studies on Otto Neurath and the Vienna Circle. Partly translated from German. 1991
                                                          ISBN 0-7923-1276-7
134. W.R. Woodward and R.S. Cohen (eds.): *World Views and Scientific Discipline Formation.* Science Studies in the [former] German Democratic Republic. Partly translated from German by W.R. Woodward. 1991
                                                          ISBN 0-7923-1286-4
135. P. Zambelli: *The Speculum Astronomiae and Its Enigma.* Astrology, Theology and Science in Albertus Magnus and His Contemporaries. 1992
                                                          ISBN 0-7923-1380-1
136. P. Petitjean, C. Jami and A.M. Moulin (eds.): *Science and Empires.* Historical Studies about Scientific Development and European Expansion.
                                                          ISBN 0-7923-1518-9
137. W.A. Wallace: *Galileo's Logic of Discovery and Proof.* The Background, Content, and Use of His Appropriated Treatises on Aristotle's *Posterior Analytics.* 1992   ISBN 0-7923-1577-4

Boston Studies in the Philosophy of Science

Also of interest:
R.S. Cohen and M.W. Wartofsky (eds.): *A Portrait of Twenty-Five Years Boston Colloquia for the Philosophy of Science, 1960-1985*. 1985   ISBN Pb 90-277-1971-3

Previous volumes are still available.

KLUWER ACADEMIC PUBLISHERS – DORDRECHT / BOSTON / LONDON

Printed in the United Kingdom
by Lightning Source UK Ltd.
104473UKS00002B/20